BASIC Statistics

BASIC statistics

J Tennant-Smith, BSc Tech, ATI, FSS

Butterworths
London Boston Durban Singapore Sydney Toronto Wellington

First published 1985

British Library Cataloguing in Publication Data

Tennant-Smith, J.
BASIC statistics.
1. Mathematical statistics — Data processing
2. Basic (Computer program language)
I. Title
519.5′028′5424 QA276.4

ISBN 0-408-01107 6

Library of Congress Cataloging in Publication Data

Tennant-Smith, J.
BASIC statistics.

Bibliography: p.
Includes index.
1. Mathematical statistics — Data processing.
2. Probabilities—Data processing. 3. Basic (Computer program language) I. Title. II. B.A.S.I.C. statistics.
QA276.4.T46 1984 519.5′028′5424 83-14474
ISBN 0-408-011107-6

Printed in England by The Thetford Press Limited, Thetford, Norfolk.

Preface

The importance of computers and computer programming today can hardly be overstated. The power of computers in science, technology, industry and commerce, is now universally recognised. Yet there are still enormous potentialities which are relatively untapped because of the lack of understanding of probability and statistics.

Specialist statisticians have made tremendous advances by using computers in applying the techniques of multivariate analysis and time series analysis. These advances usually employ computer languages such as FORTRAN, Pascal, or even more specialist mathematical languages. There remains, however, enormous scope for BASIC language programs in probability and statistics, applied to problems which are mathematically much simpler but of great practical importance.

The great advantage of BASIC is the speed with which it can be learned and applied, and the facility with which programs can be tested and corrected. Microcomputers usually employ some form of 'extended BASIC', and the further enhancements (mostly borrowed from Pascal) now offered on newer models are removing the disadvantages of BASIC for more complicated programs. This text employs the usual extended BASIC, now increasingly available on mainframe computers, but leaves it to the reader to apply any enhancements which are available on his particular model.

There are few topics of which it is more true that 'a little learning is a dangerous thing' than of probability and statistics; the concepts are sometimes rather subtle, and wrong answers are not usually 'obviously wrong' as in mathematics. Perhaps because they believe that statisticians tend to 'make a simple subject difficult', non-statisticians have shown a remarkable willingness to write textbooks in the subject. From the point of view of one whose background is industry and technology rather than mathematical statistics, it must be emphasised that (i) statistics is not an easy subject, (ii) most elementary textbooks purporting to teach statistics are absolute rubbish, (iii) most schoolteachers use bad textbooks, and (iv) many students are reluctant to abandon their misconceptions and learn the subject properly, starting from correct definitions.

In this text, careful attention has been given to *basic* definitions and concepts in probability and statistics. It is not intended that the reader should learn the subject from this text alone. He will require a sound textbook in probability and statistics to amplify the ideas, provide more examples, and give further details of formulae and techniques.

A few clues will be given in the text which should enable the reader to judge whether a textbook is sound. As a quick test, if a book fails to give clear definitions of event, independent events, discrete random variable, independent random variables, confidence interval and null hypothesis, and yet purports to teach probability and statistical techniques, it is useless.

The reader must also equip himself with the BASIC manual appropriate to the computer he is using. Chapter 1 sets out the reserved words and conventions used in the programs in this text. The reader may find it useful to produce his own notebook of 'translations' to enable him to convert programs to run on his own microcomputer, for use with microcomputer magazines as well as with this text.

Because the reader is assumed to have access to both a statistics textbook and a BASIC manual, this book concentrates on topics relevant to BASIC programming and statistics which are not usually covered in either type of book. Particular attention is given to the limitations on speed and accuracy of computers, and how to improve speed or accuracy where necessary or to produce warnings that results may be inaccurate. The reader is given an insight into the potentialities of simulation, using pseudo-random numbers, to solve practical problems.

The availability of statistical programs could increase the misuse of statistical methods. For this reason, we have tried to emphasize the correct applications of programs in the printout, such as 'X is the number of random happenings in a fixed interval of a continuum' for the Poisson distribution. We have not developed the formula for the Poisson distribution, since this is given in any good textbook on probability, but in view of the muddled presentation in many texts we have tried to destroy the myth that this distribution is concerned with 'rare events' and have emphasized the additivity of independent Poisson variables. We have similarly tried to clear up the muddle between regression on a controlled variable and the straight line relationships between two random variables.

Occasionally we have quoted from another textbook in order to point out and explain a widely-held misconception. It may be noted that all the textbooks thus criticized are listed in the Bibliography as better-than-average. Alas, most elementary textbooks in statistics are beneath criticism.

Thanks are due to colleagues at UMIST — notably Professor M. B. Priestley, Mr P. A. Wallington, and Dr P. J. Laycock — for valuable discussions in many years of teaching statistics to students of engineering and computation as well as mathematics. They take no blame for any idiosyncrasies in this text, which is a synthesis of the mathematician's approach and the technologist's approach to the topic. It should however be mentioned that the 'straight line of best fit' which is prominent in Program 10.2 and is not found in other textbooks is as much due to Dr P. J. Laycock, who recognised the relevance of the 'first principal component' in this regard, as to the author, who discovered how easily it could be programmed.

Our interest in the two-armed bandit problem which is simulated in Chapter 4 was inspired by Professor J. A. Bather, whose findings are referred to at the end of Problem 4.12. The diagram and one of the programs in Chapter 5 were originated by Mr S. Walker. Mr P. D. Smith of the Royal Military College of Science gave both general guidance and specific assistance by listing programs and testing them on a mainframe computer.

Valuable assistance also came from W. B. Smith, who made many useful comments on the first four chapters and tested the programs on a microcomputer very different from the one on which they were developed. Adrian Tennant-Smith gave advice on the reserved words and some other peculiarities of the six different types of microcomputer on which he habitually writes BASIC programs for his school-friends.

J. Tennant-Smith

Bibliography

Most microcomputers are supplied with a BASIC manual, and though no manual is beyond criticism they are usually perfectly adequate. Make full use of the manual, and improve it by your own annotations.

Books available on BASIC programming include

Alcock, D., *Illustrating BASIC,* Cambridge University Press (1977)

Forsyth, R., *The BASIC Idea,* Chapman and Hall (1978)

Gottfried, B. S., *Programming with BASIC — Schaum's Outline Series,* McGraw-Hill (1975)

Kemeny, J. G. and Kurtz, T. E., *BASIC Programming,* Wiley (1968)

Monro, D. M., *Interactive Computing with BASIC,* Arnold (1974)

For a sound presentation of elementary probability and statistics the following (though not cheap) are recommended as clear, competent, and comprehensive:

Clarke, G. M. and Cooke, D., *A Basic Course in Statistics,* Arnold (1978)

Walpole, Ronald E., *Introduction to Statistics, Third Edition,* Collier Macmillan (1982)

Of the cheaper books, better than average are

Hayslett, H. T., *Statistics Made Simple,* W. H. Allen (1968)

Hodge, S. E. and Seed, M. L., *Statistics and Probability,* Blackie (1972)

Mulholland, H. and Jones, C. R., *Fundamentals of Statistics,* Butterworths (1968)

Well worth reading for engineering and technology students interested in statistics, though we deplore their unconventional and hence confusing definition of 'significance level' or 'level of significance', are

Chatfield, C., *Statistics for Technology,* Chapman & Hall (1983)

Wetherill, G. B., *Elementary Statistical Methods,* Methuen (1967)

Contents

Chapter 1

Introduction to BASIC

1.1 Advantages and disadvantages of BASIC

BASIC (Beginner's All-purpose Symbolic Instruction Code) was first developed as an easy-to-learn, general-purpose programming language. It is a good language in which to write fairly short and simple computer programs. It is not a particularly good language for long and complicated programs.

BASIC was originally intended for use on time-sharing computer systems. Large classes of undergraduates in engineering or management sciences could all write programs in BASIC with relatively little expert guidance. Computer experts skilled in FORTRAN or Pascal have tended to sneer at BASIC because it is 'lacking in structure' — knowing very well that this phrase would be meaningless to most BASIC users. An unfortunate consequence of this snobbery is that the BASIC compilers on mainframe computers have often not been given the facilities of 'extended BASIC', which is normally available on microcomputers however modestly priced.

Microcomputers employ a BASIC interpreter rather than a compiler. This means that each line in the program is read and interpreted every time it has to be implemented in a run of the program. In contrast, a compiled program is converted into machine language after it has been written but before it is run, and it will then run much faster than an interpreted program. Since modern microcomputers are thousands of times as fast as the early mainframe computers, and since microcomputer users do not normally need extremely fast results (except in computer games, which are accordingly written in machine code), the practical disadvantages of BASIC are minimal while the ability to test and alter parts of programs very easily is a major advantage.

Hence BASIC has, rightly, become the principal language for microcomputers. All the programs in this book were developed on a microcomputer, and they employ the usual facilities of extended BASIC. But enhancements which are becoming increasingly common on newer models of microcomputer — most of them borrowed from Pascal — have not been employed in our programs.

After being developed, most of the programs were successfully run on a mainframe computer which had a fairly primitive BASIC compiler. All the programs can easily be modified, by adding a few LETs and splitting up multiple-statement lines as well as adapting one or two of the reserved words, so that they will run on a cheap home computer.

In this text we make no attempt to teach BASIC, since the reader is assumed to have access to his BASIC manual. We draw attention to the various reserved words which have been used, to some of the commoner alternatives on microcomputers, and to the conventions we have adopted in writing our programs. The reader who finds that he can improve on our programs by employing enhancements in BASIC which were not available to us, or simply by using greater wit and wisdom than we could command, is strongly encouraged to do so.

1.2 Program structure

All microcomputers have keys additional to the letters, digits, and graphics keys. For instance, there are CAPS, DELETE (or RUBOUT), SHIFT keys to move the position of the cursor, and a key which instructs the microcomputer to accept an entry and which is marked ENTER, NEWLINE, RETURN, START, or CR (presumably an abbreviation for 'carriage return'). There are also keys affecting the screen such as CLS (clear screen), HOME, SCROLL, TAB, and other single-key commands such as BREAK and possibly AUTO, COPY, ESCAPE, MODE, etc. Other commands which on most microcomputers have to be entered letter by letter are CLEAR, CONT, LIST, LOAD, NEW, RUN, SAVE, VERIFY, and possibly EDIT, OLD, and a few others. For keener users there are PEEK, POKE, and USR. All these keys and commands are outside the BASIC language (although TAB is also inside the language), are peculiar to the particular make or model of computer, and do not form part of the subject matter of this book.

In BASIC, all lines have line numbers and these determine the order of execution. A program may be modified by inputting new lines whose line numbers determine their locations in the program; if they have the same numbers as existing lines they replace those lines. (For an example, see Program 2.8 (development) in Section 2.7.) It is customary to number lines in multiples of 10 so that additional lines can readily be inserted. Some forms of BASIC will accept only one statement per line, but we have used multiple-statement lines in which statements are separated by : or \.

In our programs it is assumed that the name of any numeric

variable must consist of one letter, two letters, or a letter followed by a digit. An array is distinct from a variable even if it is given the same name, and may have one or two dimensions with up to 256 values for each dimension. The values go from 0 to a maximum of 255; some computers start the numbering in the arrays at 1, in which case minor amendments will be needed.

We make no distinction between the names of variables which are integers and the names of other numeric variables; some forms of BASIC add % to the name of an integer variable. Very large numbers and very small numbers use 'E', such as '2E+6' to represent 2 000 000 or '3E−4' to represent 0.0003. The irrational number π is assumed to be available on a key; some computers will produce π if asked for PI, but if this is not available one should simply input 3.1415927.

1.3 Types of BASIC statement

Assignments require LET on some microcomputers, but we assume that X = 1 is acceptable as an assignment. We do not use multiple assignments such as A = B = C = 0. The dimension of an array has to be declared before it is used; our microcomputer would accept DIM C(P) where the value of P has been inputted during the running of the program, but the mainframe BASIC would not accept this and so one has to allot a size which is likely to be adequate without wasting memory space. Some forms of BASIC require all numeric variables to be assigned a value before they are used in a calculation such as S1 = S1 + X; it is our usual (but not invariable) practice to 'initialize' with S1 = 0 in such a case.

Input is invited in the interactive mode by INPUT X, etc. The statement INPUT A, B, C requires the inputting of three numbers separated by commas. The alternative mode uses READ X or READ A, B, C, etc., the DATA to be read being placed anywhere in the program. On the statement RESTORE the program goes back to the beginning of the DATA for the next READ.

If the input is to be a single character, the statement GET A$ will accept the character as a string variable. Alternative versions of this statement are A$ = GET$ or A$ = INKEY$. A single-digit number may be inputted by GET X, the alternative versions being X = VAL(GET$) or X = VAL(INKEY$). If the program has to wait for this input, an IF statement must follow the GET statement (see line 50 of Program 2.2).

The user should always be told what has to be inputted at any stage. We use the word 'Input' to indicate that the carriage-return key will have to be pressed to complete the input, and the word

'Press' to indicate that the program will GET a single character or digit. In the READ. . . DATA mode it is a good practice to immediately output what has been read.

Output is by PRINT X, PRINT A$, etc., and goes to a screen or to a printer. Items to be output may be separated in the PRINT statement by a semi-colon or by a comma. Our programs assume a screen of 25 lines by 40 columns, the effect of a comma in a PRINT statement being to position the next output at whichever is next of column 0, 10, 20, or 30. TAB(8) positions the next output at column 8. We have assumed that a space is automatically inserted before any positive number. All these details of output will have to be adapted to suit the microcomputer or mainframe computer being used.

String-handling is effected by LEFT$, RIGHT$, MID$, LEN(A$), STR$, and VAL, but only one or two of our programs (notably Program 3.8) are seriously dependent on the availability of these facilities. ASC(T$) produces the code for a character, while CHR$(X) reverses this process (see lines 70 and 110 of program 2.2).

Arithmetical operators are +, –, *, /, the latter pair having priority so that 2 + 3 * 4 works out as 14 and 2–3 / 4 as 1.25. Brackets secure priority, so that (2 + 3) * 4 and 4 * (2 + 3) are both 20. Powers have overriding priority, and are denoted by ↑ or ∧ (some computers use **), so that 2*3↑2 is 2*9 which is 18. Our programs do not use DIV or MOD.

Functions assumed to be available are ABS, ATN, COS, EXP, INT, LN, LOG, SGN, SIN, SQR, TAN, though not all have been used. Note particularly that we use LN for a logarithm to base e and LOG for a logarithm to base 10. Some computers use LOG for a logarithm to base e, in which case a logarithm to base 10 is obtainable by LOG(X)/LOG(10). RND(1) produces a pseudo-random number which is equally likely to take any value between 0 and 1. We have not used DEF FN.

Conditional statements are introduced by IF. . . THEN, incorporating a relational operator between two numbers or variables. The relational operators are =, <, >, < >, < =, > =. If the relationship is true, all statements following THEN on the same line number are performed, but if it is false the program proceeds immediately to the next line number. We have not used the enhancement ELSE.

Logical operators AND and OR have been used; actually our microcomputer used * and + as logical operators, but this is so unusual that it has not been adopted in the text. We have not used the NOT operator. We have assumed that a string-equality relationship can be used, but that string inequalities cannot be used.

Program control statements include STOP and END. A loop is opened by FOR. . . TO. . . STEP. . ., but the STEP is omitted if it is unity. The loop is closed by NEXT. . . . We have not used enhancements such as REPEAT. . . UNTIL. . . or WHILE. . . WEND. PROC, for a procedure defined elsewhere, was not available to us — though we would encourage every BASIC user who finds that his computer accepts PROC to make full use of it in producing structured programs.

GOTO is of course used a great deal; the purpose of PROC is to make the tortuous GOTO meanderings which are the main drawback of BASIC unnecessary. GOSUB takes the control to a subroutine from which it will subsequently RETURN. We have used IF. . . GOTO and IF. . . GOSUB, but some versions of BASIC require THEN in such conditional statements.

ON. . . GOTO and ON. . . GOSUB have been used occasionally, but it is not difficult to adapt a program to manage without them (see note (2) to Program 5.3). REM is used for a comment, which the program ignores.

Chapter 2

Elementary descriptive statistics

2.1 What is statistics?

Statistics is a body of methods for making wise decisions in the presence of uncertainty.

Originally, 'statistics' was the study of affairs of state, and 'statists' were politicians, particularly those who kept themselves well-informed and so were able to make wise political judgements. Later the word 'statistics' was used to describe the numerical data on the basis of which statists made their judgements, and still later it was applied to numerical data in general. Although it is now very common to hear the word used simply to mean collections of numerical information, it is better to use the term 'descriptive statistics' for this purpose.

The term 'statistical inference' is used to emphasize the decision-making aspect of statistics. Statistical inference can only be effectively carried out when a certain amount of descriptive statistics and a certain amount of probability theory has been learned. The 'uncertainty' which is always present in statistical inference must be objective uncertainty; a schoolboy may be uncertain whether Richard II was the son or the grandson of Edward III, but that is not an objective uncertainty. There is objective uncertainty whether Richard III or Henry VII was responsible for the murder of the Princes in the Tower, but this would still not usually be regarded as a statistical problem because nowadays we restrict the term to numerical or quantitative problems. Often the main task of a statistician is to 'put a figure on' the uncertainty inherent in a problem, leaving the engineer or manager to make the final decision in the light of this information.

A sound text on statistics will always discuss both descriptive statistics and probability theory before embarking on statistical inference, but it is a matter of taste which of the two preliminary topics to cover first. In this text, the more complicated tasks in descriptive statistics are left until elementary probability has been dealt with; this is because the need to develop BASIC programming, beginning with short and simple programs, imposes an additional constraint.

The singular noun 'statistic' is used by statisticians to mean a numerical fact systematically collected. So the word 'statistics' can legitimately be used as a plural noun to mean numerical facts, and hence with a distinctly different meaning from 'statistics' as the body of methods for making wise decisions in the presence of uncertainty.

2.2 Types of data

Data may be *qualitative* or *quantitative.* If we observe a number of people and classify them under the headings 'male' and 'female', or by hair colour or nationality, then we are collecting qualitative data. If we record their ages, heights, or weights, then we are collecting quantitative data. Qualitative data never involves measurement, while quantitative data is often obtained by measurement. Even where quantitative data is obtained simply by observation or questioning, such as counting the number of defective light bulbs in a lighting display or asking individuals their ages in years, the results can be meaningfully represented by points on a line.

A *specimen* is an object or individual on which an *observation* or *measurement* is made. An observation is the result of observing whether or not a specimen possesses the attribute under consideration, when collecting qualitative data, but statisticians usually use the term 'observations' to include all types of data whether qualitative or quantitative.

A lecturer might classify all the students in the first year civil engineering course into 'male' and 'female', the numbers in each category being qualitative data. (Note therefore that the term 'qualitative' does not mean that there are no numbers associated with the data.)

The lecturer might then go on to collect the sizes of shoes worn by all the male students. This would be quantitative data, whether collected by measurement, observation, or questioning. Let us suppose that the results are as follows:

Data 1

9	6	9	9	9	7	6	11	6	7	6	6	10	6	7
6	8	6	5	5	5	4	6	6	7	12	6	7	8	5
10	9	7	7	5	11	9	7	6	5	7	6	5	5	12
9	8	7	9	8	5	5	6	13	11	11	5	8	10	9

A final important distinction made by statisticians is between populations and samples. A *population* is the totality of the observations with which we are concerned. A *sample* is a subset of a population. A numerical fact about a sample is termed a *statistic*, while a numerical fact about a population is termed a *parameter*.

These terms are particularly important in statistical inference. A sample is selected at random from a population, and a statistic calculated from the sample is used to draw inferences about the corresponding parameter of the population. For instance, the average shoe size of the foregoing 60 students might be used to estimate the average shoe size of all first-year civil engineering students in the country, or to test a hypothesis that the average shoe size of civil engineering students is larger than that of students in general. In either case the 60 observations would form a sample, which although not truly random would be satisfactory for drawing the inferences specified; in one case the population would be the shoe sizes of all first-year civil engineering students in the country, and in the other case it would be the shoe sizes of all civil engineering students. Care would be needed to define terms such as 'first-year', 'in the country', and 'students in general' in a clear and unambiguous manner, noting also that the inferences are confined to male students.

It might however be the case that the lecturer had no interest in statistical inference, but was solely concerned to carry out the wishes of a benefactor that each male first-year civil engineering student in the college be given a pair of shoes. In this case the 60 observations form a population. We do not always know whether a set of data is a sample or a population, and so the methods of descriptive statistics apply equally to samples or populations. It will also be noted that if the sole purpose of collecting the data is to buy pairs of shoes to fit the students then for all practical purposes the data can be regarded as qualitative rather than quantitative; there would be no interest whatsoever in the average shoe size.

2.3 Collection of qualitative data

The main task of elementary descriptive statistics in relation to qualitative data is to count the observations by type, and perhaps to sort them in some way.

Suppose that engineering products are inspected and classified into types A, B and C. The inspector could record the numbers of each type using pencil and paper, but it would be more efficient to use a computer terminal or a microcomputer. The following program would perform the task.

Program 2.1

```
10  NA = 0: NB = 0: NC = 0
20  PRINT: PRINT
30  PRINT "Input A, B or C. Input T for total."
40  INPUT T$
50  IF T$ = "A" THEN NA = NA + 1: GOTO 40
60  IF T$ = "B" THEN NB = NB + 1: GOTO 40
70  IF T$ = "C" THEN NC = NC + 1: GOTO 40
80  PRINT: PRINT "Totals are";NA;NB;NC;" out
    of";NA + NB + NC
90  PRINT: PRINT "Input Y to continue counting."
100 PRINT "Input N to start again."
110 INPUT R$
120 IF R$ = "Y" GOTO 40
130 IF R$ = "N" GOTO 10
140 GOTO 110
150 END
```

There are so many differences between types of computer and versions of BASIC that the intelligent programmer needs to learn how to adapt any BASIC program for his own use. In the early chapters of this book we shall point out various changes which may be needed, but we expect the reader to acquire the facility to adapt later programs without advice, making use of his computer manual and the indexes to the manual and to this book.

First, the program has to be made to work. Some versions of BASIC will take only one statement on each line, in which case the program must begin:

```
10  NA = 0
11  NB = 0
12  NC = 0
```

and on some computers it is also necessary to put the word LET before each of the variables NA, NB, NC in their respective lines. The lower-case letters may be difficult or impossible to use, in which case line 30 becomes: PRINT "INPUT A, B,OR C. INPUT T FOR TOTAL."

Note that in lines 50 to 70 the conditional IF applies to both the statements which follow it, so if there can only be one statement on each line we need something like the following

```
50  IF T$ = "A" GOTO 60
51  IF T$ = "B" GOTO 65
52  IF T$ = "C" GOTO 70
```

```
53  GOTO 8Ø
6Ø  NA = NA+1
61  GOTO 4Ø
65  NB = NB+1
66  GOTO 4Ø
7Ø  NC = NC+1
71  GOTO 4Ø
```

Since this is about as complicated a case of splitting up a multiple statement as we are likely to encounter, we shall assume that this lesson is now learned and (to save space) will make free use of multiple statements in future.

A final point is that some versions of BASIC will accept or require THEN in place of GOTO in lines 120 and 130, while others will require THEN GOTO.

In some forms of BASIC, lines 100 and 110 can be amalgamated as INPUT"Input N to start again"; R$. This condensed form will be used in future programs. It may also be noted that the zero can sometimes appear in programs as '0' rather than 'Ø', and for ease of typesetting the former will be used in future since its context will always distinguish it from the letter 'O'.

When the program is RUN, it will repeatedly ask for an input by printing '?'. Each entry of A, B or C will then be accepted only when the carriage return key is pressed; this key may be labelled CR, ENTER, NEWLINE, RETURN or START depending on the type of microcomputer.

If a lot of items have to be classified, it will be tedious to have to keep pressing CR. This is made unnecessary by changing line 40 to GET T$: IF T$ = "" GOTO 40. It is then a good practice to replace the word 'Input' by 'Press' in line 30 so that the user is accustomed to the fact that 'Input' means 'Press any appropriate key or keys, followed by CR' while 'Press' indicates that only a single keypress is required.

On some microcomputers the alternative to GET T$ is T$ = INKEY$. It will be noted that nothing is printed on the screen to show what key has been pressed, and it may therefore be found advantageous to insert an extra line:

```
45 PRINT T$;
```

A limitation of the GET method is that it can only be applied to single-character inputs. A single digit can be accepted by GET X, but an attempt to input a two-digit number would fail. A second drawback is that the program may GET what was intended as the previous keypress, such as a rather slow press of the CR key. A third drawback is that amendments cannot be made.

It would be easy to adapt Program 2.1 to accept labels for classes other than A, B and C provided that the different possible labels were known beforehand. The following program will classify qualitative data represented by any of the characters available on the keyboard, except that one character must be reserved as the instruction to print out the totals and '£' has been reserved for this purpose.

Program 2.2

```
 10  DIM N(255)
 20  PRINT: PRINT
 30  PRINT "Press keys for observations."
 40  PRINT "Press £ for total."
 50  GET T$: IF T$ = "" GOTO 50
 60  IF T$ = "£" GOTO 90
 70  PRINT T$;:K = ASC(T$): N(K) = N(K) + 1
 80  GOTO 50
 90  PRINT: M = 0
100  FOR I = 0 TO 255
110  IF N(I) > 0 THEN PRINT CHR$(I),N(I),: M = M + N(I)
120  NEXT I
130  PRINT: PRINT: PRINT "Total = ";M
140  PRINT: PRINT "Input Y to continue counting."
150  INPUT "Input N to start again. ";R$
160  IF R$ = "Y" GOTO 50
170  IF R$ = "N" THEN RUN
180  GOTO 140
190  END
```

This program makes use of the array N to collect the frequencies. If the possible values of I, the ASC codes of the characters which may be input, are restricted then the restriction may be inserted in lines 10 and 100. For instance, if only letters can be input then these lines become:

```
 10  DIM N(90)
100  FOR I = 65 TO 90
```

If only the digits 0 to 9 can be input, they become:

```
10 DIM N(57)
100 FOR I = 48 TO 57
```

The shoe-size data (Data 1 above) consisted of integers from 4 to 13 and could be regarded as qualitative data. Using the labels 0 to 3 to represent sizes 10 to 13 respectively, Data 1 can be inserted as input for Program 2.2. A typical line of input would be 969997616766067£. This would give a listing of the frequencies to confirm that 15 observations had been fed in. Then pressing Y and CR would allow the inputting to be continued.

The printout from Data 1 would be

```
      0     3     1     4
      2     2     3     1
      4     1     5    11
      6    14     7    10
      8     5     9     9
Total = 60
```

So the numbers of pairs required for sizes 4 to 13 respectively are 1, 11, 14, 10, 5, 9, 3, 4, 2, 1.

2.4 Sorting

The only other computing likely to be required on qualitative data is some kind of sorting. The following is a typical program for sorting as presented in microcomputer journals and manuals.

Program 2.3

```
 10  DIM A(60)
 20  PRINT: INPUT "How many values? ";N
 30  FOR I=1 TO N: INPUT A(I): NEXT I
 40  FOR I=N TO 1 STEP -1
 50  M = 0
 60  FOR J=1 TO I
 70  IF A(J)< =M GOTO 90
 80  M = A(J): L = J
 90  NEXT J
100  X = A(L): A(L) = A(I): A(I) = X
110  NEXT I
120  FOR I=1 TO N: PRINT A(I);: NEXT I
130  PRINT
140  GOTO 20
```

Even if the type of sorting here applied, known as 'exchange sorting', is accepted the program is still not very efficient. The line 70 is performed $N(N+1)/2$ times, which is 1830 times when N is 60. On about two-thirds of these occasions the conditional statement GOTO 90 will have to be performed. To find line 90, the BASIC interpreter inspects all the line numbers starting from the beginning of the program until it finds 90. So when N is 60 over 10 000 such inspections will be required. Yet a slight change in the program will eliminate the need to use a GOTO within the J loop.

There are other defects in the program, not least the fact that it will break down if there are negative values in the data.

The reader should examine the improvements in the following alternative version of lines 40 to 100:

```
 40  FOR I = N TO 2 STEP -1
 50  M = A(1): L = 1
 60  FOR J = 2 TO I
 70  IF A(J) > M THEN M = A(J): L = J
 80  NEXT J
 90  A(L) = A(I)
100  A(I) = M
```

In both versions, the program starts by finding the largest value M among the N values and its position L in the array. This value is then exchanged with the value A(N), so that the new A(N) is the largest value found. I is then set at N−1 so that the largest value among the remaining N−1 values can be found and exchanged with A(N−1). I is then reduced again, and the procedure repeated until I is 2 and the values A(1) and A(2) are put in correct order. It should be emphasized that although this is the simplest procedure for sorting numbers it is very far from being the quickest, and if N is any larger than 60 it is desirable to look for more complex but more efficient sorting methods.

Testing Program 2.3 can be rather tedious if data have to be input each time. It can be made easier by using random numbers:

```
30  FOR I = 1 TO N: A(I) = INT(100*RND(1)): NEXT I
```

This will produce N random integers between 0 and 99; in some forms of BASIC the second statement would be A(I) = INT(RND(100)) or A(I) = INT(100*RND). Provided that the number of columns is a multiple of 4, they can be printed out neatly by

```
120  FOR I = 1 TO N: IF A(I) < 10 THEN PRINT "  ";
121  PRINT A(I);: NEXT I
```

Inserting the same two lines as lines 31 and 32, followed by PRINT as line 33, will neatly print out the original integers to confirm that the sorting is correct.

When the program has been tested using random numbers, it can be converted back to input real data at line 30. Alternatively, data can be read using the READ and DATA commands. The shoe-size data in Data 1 is conveniently introduced at the end of the program.

```
 30  FOR I=1 TO N: READ A(I): NEXT I
500  DATA 9,6,9,9,9,7,6,11,6,7,6,6,10,6,7
510  DATA 6,8,6,5,5,5,4,6,6,7,12,6,7,8,5
520  DATA 10,9,7,7,5,11,9,7,6,5,7,6,5,5,12
530  DATA 9,8,7,9,8,5,5,6,13,11,11,5,8,10,9
```

Program 2.3, improved version, may now be run and will sort Data 1, but the resulting output beginning with 4 and then large numbers of the other sizes in ascending order conveys no more information than was obtained using Program 2.2. However, a simple expansion of Program 2.3 to include a new array S, in which S(I) is initially set at I but exchanged whenever A(I) is exchanged, produces a very useful tool for data analysis and presentation.

Program 2.4

```
 10  DIM A(60),S(60)
 20  PRINT: INPUT "How many values? ";N
 30  FOR I=1 TO N: READ A(I): S(I) = I: NEXT I
 40  FOR I=N TO 2 STEP -1
 50  M = A(1): L = 1
 60  FOR J=2 TO I
 70  IF A(J)>M THEN M = A(J): L = J
 80  NEXT J
 90  A(L) = A(I): A(I) = M
100  X = S(L): S(L) = S(I): S(I) = X
110  NEXT I
120  FOR I=1 TO N: PRINT S(I);A(I),: NEXT I
130  PRINT
140  GOTO 20
```

The same lines of data are added as previously, and the printout will begin:

```
22 4    35 5    40 5    52 5
44 5    21 5    57 5    20 5
```

This tells us that student number 22 in the original data takes size 4 shoes, and that size 5 shoes are worn by students 35, 40, 52, 44, 21, 57, 20, etc.

Here we have the basis for the concept of a 'data bank'. Each student is given a number, and then all relevant information on that student is filed in the appropriate location in an array. For student number I the name may be N$(I), the course of study C$(I), the home address (domicile) D$(I), and — for the purposes of our example — the shoe size could be F(I).

Suppose that it is now required to produce a list of male first-year civil engineering students in ascending order of shoe size. By examining file C$(I) and other relevant files we produce a short file S(I) containing all the numbers of the students under consideration. The statement A(I) = F(S(I)) will then produce a file of shoe-sizes corresponding to the file of student numbers, which can be sorted using Program 2.4. Since we are interested in student names rather than student numbers, the printout will be

```
FOR I=1 TO N: PRINT N$(S(I));A(I),: NEXT I
```

The idea of using an array entry as an array subscript in the form F(S(I)) or N$(S(I)) may be new to the reader, but it should cause no trouble. It can of course be avoided by first letting K = S(I) and then using K as the subscript. Note particularly that in the above procedure the data in the data bank are not moved about. The array S is used to find the sequence of student numbers in which we are at present interested, and we could of course print out D$(S(I)) as well as names and shoe-sizes if we wished to deliver the shoes to home addresses.

A possible drawback in the 'exchange sort' method is that the original sequence of the entries is subjected to unnecessary disturbance. It would be much neater if the printout from sorting Data 1 began:

```
22 4    19 5    20 5    21 5
30 5    35 5    40 5    43 5
```

so that all students taking size 5 shoes are arranged in their original sequence of student numbers. This is effected by inserting five new lines in place of lines 90–100 in program 2.4:

```
 85  IF L=I GOTO 110
 90  X = S(L)
 95  FOR J=L TO I-1
100  A(J) = A(J+1): S(J) = S(J+1): NEXT J
105  A(I) = M: S(I) = X
```

It will be seen that in this 'tidy sorting' the values between L and I are all moved down instead of merely interchanging A(L) and S(L) with A(I) and S(I), and so sorting takes longer. Line 85 is essential, since in BASIC the FOR loop is always executed once even when the condition which is tested when NEXT is reached is never satisfied. A run of this modified program will show that it is also necessary to change > in line 70 to >=.

The most efficient way to sort data on a microcomputer is to POKE the data into the memory and then use a machine-code program for the sorting loop constituting lines 40–110 of Program 2.4. Such a program will carry out a 'tidy sort' at least thirty times as fast as the BASIC program, and hence faster than most other programming languages can achieve without using machine-code. So the best advice to BASIC programmers encountering the drawbacks of BASIC is to learn another programming language if the problem is complexity in the program, but to learn machine-code programming if the problem is slowness in running.

The final program in this section should, with the help of the REM statements and the foregoing discussion, be self-explanatory. Note that only a 'tidy sort' method will work. The reader may care to add a lot more words to the data, especially four-lettered words, to confirm that the sorting is always effective but very slow as N becomes large.

Program 2.5

```
  5 REM PROGRAM TO SORT UP TO 60 WORDS ALPHA-
    BETICALLY — VERY SLOWLY!
 10 DIM  A$(60)
 20 PRINT: INPUT "How many words? ";N
 30 LW = 0
 35 REM READ THE WORDS, AND FIND THE LENGTH LW
    OF THE LONGEST WORD.
 40 FOR I=1 TO N
 50 READ W$: IF LEN(W$) >LW THEN LW=LEN(W$)
 60 A$(I) = W$
 70 NEXT I
 75 REM EXTEND ALL WORDS TO LENGTH LW BY
    ADDING ZEROES.
 80 FOR I=1 TO N
 90 W$ = A$(I)
100 FOR J = LEN(W$)+1 TO LW: W$ = W$+"0": NEXT J
110 A$(I) = W$
120 NEXT I
125 REM APPLY TIDY SORT TO EACH LETTER POSITION,
    WORKING BACKWARDS.
130 FOR H = LW TO 1 STEP -1
140 FOR I = N TO 2 STEP -1
150 M = ASC(MID$(A$(1),H,1)): L = 1
160 FOR J=2 TO I
170 K = ASC(MID$(A$(J),H,1))
180 IF K> =M THEN M = K: L = J
190 NEXT J
```

```
200 IF L=I GOTO 240
210 W$ = A$(L)
220 FOR J=L TO I–1: A$(J) = A$(J+1): NEXT J
230 A$(I) = W$
240 NEXT I
250 NEXT H
255 REM THE PRINTOUT INCLUDES THE REDUNDANT
    ZEROES.
256 REM THE CONSCIENTIOUS READER WILL WANT TO
    IMPROVE ON THIS.
260 FOR I=1 TO N: PRINT A$(I),: NEXT I
270 RESTORE
280 GOTO 20
500 DATA JONES, BROWNE, KAYE, SMITH, LEE, HALL
510 DATA PRICE, KAY, BROWN, WRIGHT, AARON, LAW
```

Because a GOTO statement consumes a lot of running time if it is in an oft-repeated loop, an improvement is possible in those versions of BASIC which permit multiple statements governed by the same IF. Delete line 200, and change line 220 to

```
220 IF L<I THEN FOR J=L TO I–1: A$(J) = A$(J+1): NEXT J
```

and also incorporate lines 210 and 230 into the IF statement if there is room.

2.5 Collection of quantitative data

Program 2.2 was used to collect Data 1 by frequencies, treating it as qualitative data. Relatively few changes are needed to produce a program for collecting quantitative data, provided that the observations are all single digits.

Program 2.6

```
 10 DIM N(9)
 20 PRINT: PRINT
 30 PRINT "Press keys for observations."
 40 PRINT "Press £ for total."
 50 GET T$: IF T$="" GOTO 50
 60 IF T$="£" GOTO 90
 70 PRINT T$;: K=ASC(T$)–48: IF K<0 OR K>9 GOTO 50
 80 N(K) = N(K)+1: GOTO 50
 90 PRINT: N = 0
100 FOR I=0 TO 9
```

```
110 M = N(I): N = N+M: PRINT I,M
120 NEXT I
130 PRINT: PRINT N;" observations"
140 PRINT: PRINT "Input Y to continue counting."
150 INPUT "Input N to start again. ";R$
160 IF R$="Y" GOTO 50
170 IF R$="N" THEN RUN
180 GOTO 140
190 END
```

This program has incorporated a test in line 70 so that a keypress which is not a digit simply reverts to line 50 rather than causing a break in the running of the program. In one type of microcomputer this line is best written as

```
70 PRINT T$;: K=ASC(T$)-48: IF (K<0)+(K>9) THEN
   USR(62): GOTO 50
```

reflecting a different form of the OR logical operator, and using a built-in 'bleep' which on some microcomputers is obtained by PRINT CHR$(7).

We can conveniently produce a set of invented single-digit data by taking Data 1 and subtracting 10 from each observation which is greater than 9:

Data 2

9	6	9	9	9	7	6	1	6	7	6	6	0	6	7
6	8	6	5	5	5	4	6	6	7	2	6	7	8	5
0	9	7	7	5	1	9	7	6	5	7	6	5	5	2
9	8	7	9	8	5	5	6	3	1	1	5	8	0	9

The printout will be exactly the same as when Data 1 was collected using the digits 0 to 3 to represent sizes 10 to 13. But in the case of quantitative data we are likely to be interested also in cumulative frequencies, which would have no meaning for qualitative data. For any value x, the *cumulative frequency* is the number of observations which are less than or equal to x.

To obtain cumulative frequencies from Program 2.6 it is only necessary to change the last statement on line 110 to PRINT I,M,N. The printout when using Data 2 will then be

```
0    3     3
1    4     7
2    2     9
3    1    10
4    1    11
```

5	11	22
6	14	36
7	10	46
8	5	51
9	9	60

although to get the numbers in columns like this it would be necessary to replace line 110 by

```
110 M = N(I): N = N+M: PRINT I,
112 IF M<10 THEN PRINT " ";
114 PRINT M,
116 IF N<10 THEN PRINT " ";
118 PRINT N
```

The graphical presentation of this kind of data is best achieved by a *vertical bar chart,* representing the frequency of each value by a vertical bar of the appropriate length. Provided that the total number of observations is not too large, a reasonably good chart can be obtained by adding a few lines and a subroutine to Program 2.6.

Program 2.6 (extension)

```
 95 MX = 0
115 IF M>MX THEN MX = M
135 GOSUB 200
200 PRINT: PRINT
210 FOR I=MX TO 1 STEP -1
220 IF I<10 THEN PRINT " ";
230 PRINT I;" ";
240 FOR J=0 TO 9
250 IF N(J) > =I THEN PRINT":  ";
260 IF N(J)<I THEN PRINT"   ";
270 NEXT J
280 PRINT
290 NEXT I
300 PRINT TAB (3);: FOR I=0 TO 9: PRINT I;" ";: NEXT I
310 PRINT: RETURN
```

The enterprising reader will have little difficulty in improving on this subroutine to make better use of the available graphics, and perhaps to adapt it so that larger frequencies can be represented than the number of lines of print in the display. It is also desirable to make provision for halting the display when the frequencies and cumulative frequencies have been printed and requiring a further keypress before the vertical bar chart is printed.

Cumulative frequencies are represented graphically by a *cumulative frequency diagram*. It will be seen from the definition of cumulative frequencies that for Data 2 they take the value zero for x less than 0, 3 for $0 \leqslant x < 1$, 7 for $1 \leqslant x < 2$, and so on, the cumulative frequency being 60 for all $x \geqslant 9$. So the cumulative frequency diagram consists of a lot of horizontal lines, termed a 'step function'. It is left to the reader to develop the graphics needed to represent this function.

There is a famous batch of real data which can be collected using Program 2.6, but would require development of the subroutine in order to represent the frequencies graphically since the value 0 occurs no less than 144 times. In 1898 von Bortkiewicz collected data on the numbers of cavalrymen killed by horse-kicks in each of the 20 years from 1875 to 1894 for each of 14 army corps. These data have been studied to see whether they suggest that such deaths were perfectly random or whether certain army corps were particularly accident-prone; but our present interest is only in descriptive statistics.

Data 3

Corps	*1*	*2*	*3*	*4*	*5*	*6*	*7*	*8*	*9*	*10*	*11*	*12*	*13*	*14*
1875	0	0	0	0	0	0	0	1	1	0	0	0	1	0
1876	2	0	0	0	1	0	0	0	0	0	0	0	1	1
1877	2	0	0	0	0	0	1	1	0	0	1	0	2	0
1878	1	2	2	1	1	0	0	0	0	0	1	0	1	0
1879	0	0	0	1	1	2	2	0	1	0	0	2	1	0
1880	0	3	2	1	1	1	0	0	0	2	1	4	3	0
1881	1	0	0	2	1	0	0	1	0	1	0	0	0	0
1882	1	2	0	0	0	0	1	0	1	1	2	1	4	1
1883	0	0	1	2	0	1	2	1	0	1	0	3	0	0
1884	3	0	1	0	0	0	0	1	0	0	2	0	1	1
1885	0	0	0	0	0	0	1	0	0	2	0	1	0	1
1886	2	1	0	0	1	1	1	0	0	1	0	1	3	0
1887	1	1	2	1	0	0	3	2	1	1	0	1	2	0
1888	0	1	1	0	0	1	1	0	0	0	0	1	1	0
1889	0	0	1	1	0	1	1	0	0	1	2	2	0	2
1890	1	2	0	2	0	1	1	2	0	2	1	1	2	2
1891	0	0	0	1	1	1	0	1	1	0	3	3	1	0
1892	1	3	2	0	1	1	3	0	1	1	0	1	1	0
1893	0	1	0	0	0	1	0	2	0	0	1	3	0	0
1894	1	0	0	0	0	0	0	0	1	0	1	1	0	0

The printout for Data 3 from Program 2.6 is

```
0     144     144
1      91     235
2      32     267
3      11     278
4       2     280
```

with a frequency of zero and cumulative frequency 280 for all values higher than 4.

If the data do not consist of single digits, a simple program for collecting the data requires the number of observations and the minimum and maximum possible values to be known before the data are input. The following program will then collect the data and print out the frequencies and cumulative frequencies.

Program 2.7

```
 10 PRINT: PRINT
 20 INPUT "Input number of observations ";N
 30 INPUT "Input minimum and maximum values ";MN,MX
 40 PRINT: R=MX–MN: DIM N(R)
 50 FOR I=1 TO N
 60 PRINT "Input observation No.";I;": ";
 70 INPUT X: N(X–MN) = N(X–MN)+1
 80 NEXT I
 90 N = 0
100 FOR I=0 TO R
110 IF N(I)>0 THEN N = N+N(I): PRINT MN+I,N(I),N
120 NEXT I
130 PRINT: PRINT N;" observations"
140 PRINT: INPUT "Input Y for a new set of data. ";R$
150 IF R$="Y" THEN RUN
160 END
```

We try to keep our programs simple, and leave it to the reader to make them more foolproof if desired. A typical addition to Program 2.7 for this purpose would be

```
75 IF X<MN OR X>MX GOTO 60
```

2.6 Measures of location ('Averages')

The most important statistic to be derived from a set of quantitative data is usually an 'average' or 'measure of location'; that is, a statistic which indicates where the 'centre' of the data lies.

Just as people may have different opinions as to where the centre of a city is, depending on what they would want to do when they get to the centre, so there are different ways of assessing the average of a set of data. Examples of types of average, each of which has a strict specification, are the arithmetic mean, the mid-range, the median, the mode, the geometric mean, and the harmonic mean.

All types of average have one simple property in common. An average always takes a value not less than the minimum value observed and not greater than the maximum value observed. It follows that if all the observations take the same value then all the different types of average must also take that same value.

Three types of average can be read off so easily from the collected data that it is hardly necessary to incorporate them into a statistical program. The *mid-range* is defined as the value which is halfway between the minimum and the maximum values observed. The *median* is the middle value when the observations are arranged in increasing order of magnitude; if the number of observations is even there will be two middle values, and the median lies halfway between these. The *mode* is the most frequently occurring value; in some cases there are two or more modes with the same frequency.

In the case of Data 2 the mid-range is 4.5, the median is 6 (since the 30th and 31st values are both equal to 6), and the mode is also 6 since its frequency 14 is greater than any other frequency.

In the case of Data 3 the mid-range is 2, the median is 0, and the mode is also 0. Data 4 provides a case where the mode differs from the median.

Data 4

38, 50, 37, 44, 41, 53, 42, 48, 43, 42, 46

For Data 4 the minimum is 37 and the maximum is 53, and so the mid-range is 45. It will be found that 5 of the 11 values are less than 43 and that 5 are greater than 43, and so the median is 43. The only value which occurs more than once is 42, and so the mode is 42.

Requiring rather more calculation than the three measures of location already considered, the *arithmetic mean* is nevertheless the most commonly used average since it takes all the values fully into account. Often termed simply the *mean,* it is defined as the sum of the observations divided by the number of observations. Program 2.7 can be adapted to calculate and print out the mean by making small additions to three of the lines:

```
90 N = 0: S1 = 0
110 IF N(I)>0 THEN N = N+N(I): X = MN+I: PRINT
    X,N(I),N: S1 = S1+X*N(I)
130 PRINT: PRINT N;" observations": PRINT " Mean
    =";S1/N
```

It will be found that the mean of Data 2 is 5.75, the mean of Data 3 is 0.70, and the mean of Data 4 is 44.0.

There are three special types of average which are obtained by calculating the arithmetic mean of a specified transformation of the observed values and then reversing the transformation. The *geometric mean* is the antilogarithm of the arithmetic mean of the logarithms of the observations. The *harmonic mean* is the reciprocal of the arithmetic mean of the reciprocals of the observations. The *root mean square* is the squareroot of the arithmetic mean of the squares of the observations.

These special means should be used only when they are known to be appropriate to the type of data to which they are applied; they should certainly not be presented merely for the sake of variety. As a rough guide, one can expect to use the geometric mean when averaging a sequence of ratios; to use the harmonic mean when averaging a sequence of velocities over equal distances (whereas velocities over equal periods of time would require the arithmetic mean); and to use the root mean square in certain problems involving moments of inertia or radii of gyration.

An alternative definition of the geometric mean is the Nth root of the product of the N observations, but this does not offer any advantage in BASIC programming. It should be noted that if any of the observations is negative or zero then the geometric mean and the harmonic mean do not exist.

The special means may be incorporated into Program 2.7 in a similar way to that proposed for the arithmetic mean. The sets of required changes are

Line 90	*Line 110*	*Line 130*
S1 = 0	S1 = S1 + X*N(I)	PRINT "Mean = " ;S1/N
SG = 0	SG = SG + N(I)*LN(X)	PRINT " Geometric Mean =";EXP(SG/N)
SH = 0	SH = SH + N(I)/X	PRINT " Harmonic Mean =";N/SH
S2 = 0	S2 = S2 + X*X*N(I)	PRINT " Root Mean Square =";SQR(S2/N)

It may be useful to have a program which will calculate a variety of measures of location, even though they will never all be useful for the same set of data. The median and mode are easily read off from the frequency data and are rather difficult to include in a simple program, but Program 2.8 calculates the other five of the measures we have considered. Two other features are that it is now assumed that there is no requirement to collect the data by frequencies, and it is assumed that the number of observations is not known in advance and so the end of the data is indicated by inputting a very large number.

Program 2.8

```
 10 PRINT: PRINT " MEASURES OF LOCATION"
 20 PRINT: PRINT "Input observations."
 30 PRINT "Input 1E+11 to end input."
 40 MN = 1E+11: MX = -MN
 50 N = 0: S1 = 0
 60 SG = 0: SH = 0: S2 = 0
 70 INPUT X: IF X>1E+10 GOTO 130
 80 IF X<MN THEN MN = X
 90 IF X>MX THEN MX = X
100 N = N+1: S1 = S1+X
110 SG = SG+LN(X): SH = SH+1/X: S2 = S2+X*X
120 GOTO 70
130 PRINT: PRINT N;" observations"
140 PRINT: PRINT "Mid-range = ";(MN+MX)/2
150 PRINT "Arithmetic Mean = ";S1/N
160 PRINT "Geometric Mean = ";EXP(SG/N)
170 PRINT "Harmonic Mean = ";N/SH
180 PRINT "Root Mean Square = ";SQR(S2/N)
190 PRINT
200 GOTO 20
```

In some cases the data which are to be processed are already collected by frequencies, either because they originated in this form or because they have been collected by an operation such as that in Program 2.6. By changing three lines, Program 2.8 can be adapted so that it accepts data in this form:

```
 20 PRINT: PRINT "Input values with their frequencies."
100 INPUT "Frequency: ";F: N = N+F: S1 = S1+F*X
110 SG = SG+F*LN(X): SH = SH+F/X: S2 = S2+F*X*X
```

The frequency data for Data 2 and Data 3 have already been given above. Two further sets of such data are

Data 5

Values:	10	12	14	16	17	19	20	21
Frequencies:	2	3	4	5	6	8	10	12

Data 6

Values:	15	16	17	18	19	20	21	22	23	24	25
Frequencies:	1	3	6	9	12	14	11	9	5	0	1

In Data 5 we have listed only the values with non-zero frequencies while in Data 6 we have listed all the values from the minimum to the maximum. Either convention may be encountered in practice.

For the benefit of the enthusiastic reader who wishes to check his programs we list all seven measures of location for the five sets of quantitative data so far discussed.

Type of average	*Data 2*	*Data 3*	*Data 4*	*Data 5*	*Data 6*
Mid-range	4.5	2	45	15.5	20
Median	6	0	43	19	20
Mode	6	0	42	21	20
Arithmetic mean	5.75	0.70	44.0	17.96	19.77
Geometric mean	—	—	43.8	17.65	19.67
Harmonic mean	—	—	43.5	17.28	19.56
Root mean square	6.26	1.12	44.2	18.23	19.88

Data 2 and *Data 3* include zeroes and so neither the geometric mean nor the harmonic mean exist.

Common sense should dictate how many decimal places appear in the quoted statistics. For instance, only one decimal place is quoted for calculated averages for Data 4 because there are only 11 observations which are all integers.

Programs 2.7 and 2.8 can of course also accept non-integer observations, provided that in the case of Program 2.8 the observations are all positive. An example of such data is

Data 7

3.1, 4.2, 5.5, 3.7, 5.5, 7.0, 5.2

For Data 7, the seven averages work out as 5.05, 5.2, 5.5, 4.89, 4.73, 4.58, 5.04 respectively.

Programs to deal with data which require to be grouped will be considered in Chapter 5. At that stage we shall also indicate how a comprehensive 'descriptive statistics package' could be developed, instead of writing a lot of short programs each with a limited objective.

2.7 Measures of dispersion

In some cases the only statistic which is of importance is a measure of location, but in many other cases it is necessary also to have a measure of dispersion (also called 'spread' or 'variation'). If we measured 50 bearings which were required to be 18 mm in diameter and found that the mean diameter was 17.96 mm, we would not remain satisfied for long if the detailed measurements proved to be those given as Data 5. Without needing the details, however, we could pinpoint what was wrong by being given a measure of dispersion.

The simplest measure of dispersion is the *range,* which is the difference between the minimum and the maximum values observed. But the range of a sample of 12 observations from a large population would be a much less satisfactory guide to the dispersion of the whole population than could be obtained using an alternative measure which took all the values fully into account instead of looking only at the extreme values.

The most obvious way to measure dispersion taking all values into account is to take the mean of the 'deviations', that is, the differences between the observations and the arithmetic mean. If the N observations are denoted by $x_1, x_2 \ldots, x_N$, and their arithmetic mean is denoted by $\bar{x}$, then

$$\text{Arithmetic mean } \bar{x} = \frac{x_1 + x_2 + \ldots + x_N}{N} = \frac{1}{N}\sum_{i=1}^{N} x_i$$

$$\text{Mean deviation } = \frac{1}{N}\sum_{i=1}^{N} \left| x_i - \bar{x} \right|$$

The vertical lines indicate that we take the positive value of the difference between x_i and $\bar{x}$, irrespective of which is the larger. For emphasis the mean deviation is sometimes called the *mean absolute deviation,* but there can be no ambiguity since the mean of $(x_i - \bar{x})$ retaining the negative sign where applicable would always prove to be zero.

There are good reasons for insisting that if deviations are to be measured from a particular 'centre' of the data and then averaged, the 'centre' should be chosen so as to make the average deviation as small as possible. It can be shown that the mean deviation from the median is always less than or equal to the mean deviation from the mean, and so the former is to be preferred.

A simple program to find the mean deviation requires that the mean or median should already have been found.

Program 2.9

```
 10 PRINT: PRINT
 20 INPUT "Input number of observations ";N
 30 INPUT "Input mean and median ";M,D
 40 PRINT: T1 = 0: T2 = 0
 50 FOR I=1 TO N
 60 PRINT "Input observation No.";I;": ";
 70 INPUT X: T1 = T1+ABS(X-M): T2 = T2+ABS(X-D)
 80 NEXT I
 90 PRINT: PRINT N;" observations"
100 PRINT: PRINT "Mean deviation about the mean =";T1/N
```

```
110 PRINT "Mean deviation about the median =";T2/N
120 GOTO 10
```

By changing four lines, this program can be adapted to accept data collected by frequencies:

```
20 INPUT "Input number of different values ";K
50 N = 0: PRINT "Input values with their frequencies ": PRINT
60 FOR I=1 TO K
70 INPUT X,F: N = N+F: T1 = T1+F*ABS(X-M):
   T2 = T2+F*ABS(X-D)
```

Three additional lines would permit a check to be made on the value ascribed to the mean:

```
45 CM = 0
75 CM = CM+F*(X-M)
105 PRINT "Using mean =";M;" the error = "; CM: PRINT
```

Checking the value ascribed to the median would be rather more difficult to program.

In practice the use of absolute values makes the mathematical treatment of the mean deviation awkward, and it is much more convenient to use the squares of deviations as the basis for a measure of dispersion. This is because it can be shown that

$$\sum_{i=1}^{N}(x_i-\bar{x})^2 = \sum_{i=1}^{N}x_i^2-N\bar{x}^2$$

It is accordingly possible to compute the sum of squared deviations by a single input of the data, instead of having to determine the mean (or median) and then look at the individual observations again. A further advantage is that, by the same principle which makes the median the appropriate 'centre' when computing the mean of the absolute deviations, the arithmetic mean is the appropriate 'centre' when computing the mean of the squared deviations; i.e., it makes the mean squared deviation a minimum.

The mean squared deviation is sometimes called the 'variance', but it is more usual to reserve this term for a quantity which is $N/(N-1)$ times the mean squared deviation:

$$\text{Variance} = s^2 = \frac{1}{N-1}\sum_{i=1}^{N}(x_i-\bar{x})^2 = \frac{1}{N-1}\left(\sum_{i=1}^{N}x_i^2-N\bar{x}^2\right)$$

This is because the mean squared deviation of small samples tends to understate the mean squared deviation of the populations from which they are drawn, the extreme case being samples of size 1 whose mean squared deviation is always zero. It can be shown that the variance of a sample always gives an 'unbiased estimate' of the variance of the population from which it is drawn; i.e., it neither tends to understate it nor tends to overstate it. The variance of a sample of size 1 is zero divided by zero and so is indeterminate.

The square root of the variance is also useful, being measured in the same dimensions as the original observations and the arithmetic mean, and it is termed the *standard deviation.* It is customary to denote the standard deviation by s, and so simplest to denote the variance by s^2.

Program 2.8 can be amended and extended to produce measures of dispersion.

Program 2.8 (extension)

```
 10 PRINT: PRINT " MEASURES OF LOCATION AND
    DISPERSION"
150 M = S1/N: PRINT "Arithmetic Mean = ";M
190 PRINT: PRINT "Range = ";MX-MN
200 SS = S2-M*S1: M2 = SS/N: PRINT "Mean Square
    Deviation = ";M2
210 PRINT "Root Mean Square Deviation = ";SQR(M2)
220 V = SS/(N-1): PRINT "Variance = ";V
230 PRINT "Standard Deviation = ";SQR(V)
240 PRINT
250 GOTO 20
```

It will be noted that if S1 is the sum of the observations and M is the mean then M = S1/N and so N M^2 (alternatively written as $N\bar{x}^2$) is most simply computed as M*S1.

For strictly positive data, such as measurements of natural phenomena or of manufactured products, it is sometimes useful to quote the *coefficient of variation* which is the standard deviation divided by the mean. The coefficient of variation is dimensionless; for instance, the heights of adult males of common ethnic origin have a coefficient of variation of approximately 0.06 irrespective of the units in which the heights are measured.

The following new lines will introduce the coefficient of variation into the foregoing program:

```
230 SD = SQR(V): PRINT "Standard Deviation = ";SD
240 PRINT "Coefficient of Variation ";: IF M>0 THEN PRINT
    " = ";SD/M
250 IF M< =0 THEN PRINT "does not exist."
260 PRINT: GOTO 20
```

However, if the possibility of non-positive values has to be accepted it is necessary to make other changes to Program 2.8. The following lines introduce this possibility, and also the possibility of data collected by frequencies, by asking two questions about the data. Note the widely-used convention in programming microcomputers that '(Y/N)' is shorthand for 'Input Y if the answer is yes; input N if the answer is no'.

Program 2.8 (development)

```
 20 PRINT: INPUT "Is the data strictly positive? (Y/N) ";P$
 22 PRINT: INPUT "Is the data collected by frequencies? (Y/N)
    ";F$
 24 PRINT: IF F$ = "Y" THEN PRINT "Input values with their
    frequencies."
 26 IF F$ = "N" THEN PRINT "Input observations."
 28 PRINT: F = 1
100 IF F$ = "Y" THEN INPUT "Frequency: ";F
105 N = N + F: S1 = S1 + F*X
110 IF P$ = "Y" THEN SG = SG + F*LN(X): SH = SH + F/X
115 S2 = S2 + F*X*X
160 PRINT "Geometric Mean ";: IF P$ = "Y" THEN PRINT
    " = ";EXP(SG/N)
165 IF P$ = "N" THEN PRINT "does not exist."
170 PRINT "Harmonic Mean ";: IF P$ = "Y" THEN PRINT
    " = ";N/SH
175 IF P$ = "N" THEN PRINT "does not exist."
230 SD = SQR(V): PRINT "Standard Deviation = ";SD
240 PRINT "Coefficient of Variation ";: IF P$ = "Y" THEN
    PRINT " = ";SD/M
250 IF P$ = "N" THEN PRINT "does not exist."
260 PRINT: GOTO 20
```

If Program 2.8 is treated to the 'extension' as far as line 220, and then to the 'development', a comprehensive program is obtained for calculating most of the measures of location and dispersion on data which are either in single observations or collected by frequencies. A still more comprehensive program will be developed in Chapter 5, when the handling of grouped data will be discussed and some other statistical terms such as semi-interquartile range, skewness, and kurtosis will be introduced.

It only remains for the enthusiastic reader to run the six sets of quantitative data through the developed program and to check the results against the previous list of averages and the following list of measures of dispersion.

Measure of dispersion	*Data 2*	*Data 3*	*Data 4*	*Data 5*	*Data 6*	*Data 7*
Range	9	4	16	11	10	3.9
Mean deviation about the mean	1.88	0.720	3.82	2.61	1.63	1.04
Mean deviation about the median	1.82	0.700	3.73	2.40	1.61	1.00
Mean square deviation	6.19	0.760	21.8	9.64	4.06	1.48
RMS deviation	2.49	0.872	4.67	3.10	2.02	1.22
Variance	6.29	0.763	24.0	9.84	4.12	1.73
Standard deviation	2.51	0.873	4.90	3.14	2.03	1.32
Coefficient of Variation	—	—	0.111	0.175	0.103	0.269

PROBLEMS

(2.1) Modify Program 2.1 so that it will count observations classified as A, B, C, D, E, making any necessary changes in the BASIC so that the program is acceptable to your computer. Start a notebook of such changes for future reference. Run the program, inputting 30 invented observations. Modify the program at lines 30 and 40 so that it will GET a single-key entry, and run the modified program.

(2.2) Modify Program 2.2 so that it will accept only letters, declaring DIM N(25) at line 10. (You will need to study your character codes to ensure that an array of size only 26 is sufficient.) Run the program so that it counts the letters of at least 20 words of your favourite poem. Replace line 70 by two or more lines to ensure that if a key is pressed which is not a letter the program will (after a bleep if possible) revert to line 50 instead of declaring a DATA ERROR.

(2.3) Run 30 numbers through Program 2.3 to make sure that it sorts them correctly, both in the original and in the improved version. Modify it so that it generates random numbers at line 30 and run it again.

(2.4) If you are using a microcomputer with a timing device, modify Programs 2.3 and 2.4 by starting the device with a new line 35 and measuring the time with a new line 115. Time the programs when N = 30 and when N = 60. Also time Program 2.4 as modified for 'tidy sorting', for these values of N and for much larger values of N.

(2.5) If you are using a microcomputer with a timing device, modify Program 2.5 by starting the device at line 126 and measuring the time at line 251. Add further lines of DATA, and time the sorting of 30 words and of 60 words. Note that the times increase more than proportionately to the numbers of words.

(2.6) Try to improve Programs 2.4 and 2.5 in all ways which you can. In particular, expunge the redundant zeroes in the printout of Program 2.5.

(2.7) Run all the programs in Sections 2.5 to 2.7 and check all the measures of location and dispersion for Data 2 to Data 7.

(2.8) Try to amalgamate the programs in Sections 2.5 to 2.7 into a large comprehensive program.

(2.9) Calculate all the measures of location and dispersion for the following sets of data, using your programs. Keep a record of the results, for comparison with the results obtained in Chapter 5.

Crushing loads data

The following are (allegedly) the crushing loads of 48 mortar cubes in pounds per square inch:

4596	5196	5329	5249	5542	5349	5495	4916
4829	5176	5142	5282	5129	5442	4956	5522
4723	5355	5349	4883	5029	5522	5249	5482
5082	5435	5176	5349	5349	4709	5062	5236
4509	5482	4989	5575	5189	4922	5702	4989
5036	5142	5089	5182	5216	4676	5329	5309

Metal shaft data

The following are (allegedly) the diameters in inches of 30 metal shafts:

2.989	3.042	3.016	3.022	3.041	2.998	3.016	3.014	3.041	3.051
3.017	3.028	3.046	2.997	3.013	3.024	3.019	2.999	3.016	3.052
2.998	3.002	3.017	3.026	3.019	2.994	3.001	3.024	3.027	3.006

(2.10) If you are using a microcomputer with good 'graphics', try to develop Program 2.6 (extension) so that it produces a cumulative frequency diagram as described in Section 2.5.

Chapter 3

Speed and accuracy of computation

3.1 Limitations in speed and accuracy

Both the friends and the enemies of computers often regard them as all-powerful, as if no number is too large for computers to handle, no calculation so complex that it cannot be handled in a few seconds, and for all practical purposes the accuracy of computers is beyond question or reproach. All these views are false. Computers are severely limited in the sizes of number they can handle, in their speed of calculation, and in the accuracy of their results.

All computer users should be aware of these limitations, and should learn how to make the best of their computers. But this need is particularly important for those who wish to acquire some modest expertise in using computers to apply statistical techniques. It is even more important for those who intend to write their programs in BASIC, because there are (at the time of writing) no universally accepted standards of accuracy for BASIC as there are for the main 'scientific' computer languages. Some manufacturers of microcomputers and authors of BASIC software appear to believe that 4-figure or 5-figure accuracy is good enough for anybody, and if their usual customers are interested only in good graphics and colourful computer games they are entitled to their beliefs. It is up to the scientific user to test his computing equipment to ensure that it is fit for his purpose, just as he would insist on using a properly calibrated thermometer rather than any old garden thermometer if he needed to make accurate determinations of temperature.

Try running the following program:

Program 3.1

```
10 X = 1
20 X = X/2: PRINT X: IF X>0 GOTO 20
30 END
```

For a typical microcomputer the printout ends:

```
.54210109E-19
.27105054E-19
0
```

So this microcomputer cannot handle a quantity smaller than 0.27×10^{-19}. After working repeatedly with about 8-figure accuracy it suddenly decides that $.13552527 \times 10^{-19}$ is indistinguishable from zero!

Now try the following program:

Program 3.2

```
10 Y = 1
20 X = Y: Y = 2*X: PRINT Y: GOTO 20
30 END
```

In this case the printout ended by declaring a DATA ERROR when it tried to multiply 0.4611686E+19 by 2. So the microcomputer cannot handle a number greater than 0.46×10^{19}, but at least it draws attention to the fact rather than making erroneous calculations.

The third test program is:

Program 3.3

```
10 X = 1
20 X = X/2: Y = X+1-1: PRINT X,Y: IF Y>0 GOTO 20
30 END
```

The last two lines of printout are

```
.29802322E-07    .29802322E-07
.14901161E-07    0
```

So the microcomputer completely loses the value 0.149×10^{-7} if it has to add 1 to it and then subtract 1 again, even though this value is a million million times as great as the smallest value the computer can hold! A similar effect arises when using large numbers:

Program 3.4

```
10 X = 1
20 X = 2*X: Y = X+1: PRINT X,Y: IF Y>X GOTO 20
30 END
```

The printout ceases to show the difference between X and X+1 when X is greater than 10^8, since only eight digits are printed. The program terminates after printing .42949673E+10, showing that our microcomputer has the same limitation as is explicitly stated in the manual for a different model: 'The largest integer (whole number) that can be held completely accurately is $2^{32}-1$ (4 294 967 295)'.

It may be worthwhile to test variations of the condition in line 20 in Program 3.4, such as IF Y−X=1 or IF Y=X+1. Statements which are 'obviously' true or equivalent in elementary arithmetic may not be treated as such by the computer when it is at the limit of its capacity for accurate calculation.

The reader should bear in mind the limitations of the BASIC he is using when applying the programs in this book. It is our purpose to draw attention to the difficulties as well as to the opportunities in BASIC programming, but not to make all our programs foolproof — which would require them to be much longer and more complicated.

3.2 Speed tests

There are three ways of making sure that BASIC programs run as fast as they can: deducing what statements will run quickest from knowledge of how the BASIC interpreter works, picking up tips from manuals or other sources, and carrying out one's own speed tests.

In this section we offer the results of a few speed tests, but also show how the reader may make his own tests provided that he is using a computer with some kind of timing device. Program 3.5 is the basis of our speed tests, carried out on a microcomputer.

Program 3.5

```
 10 INPUT "SPEED TEST: Input length of run ";N
 20 REM INSERT PROGRAM LINE OR LINES HERE TO SET
    TIMER TO ZERO.
 30 FOR I = 1 TO N
 40 X = RND(1)
 60 NEXT I
 70 REM INSERT PROGRAM LINE OR LINES HERE TO FIND
    T = TIME IN SECS.
 80 PRINT N,T
 90 GOTO 10
100 END
110 REM LINE 150 BECOMES LINE 50 FOR TEST OF
    METHOD 1.
150 Y = X↑2
```

```
210 REM LINE 250 BECOMES LINE 50 FOR TEST OF
    METHOD 2.
250 Y = X*X
```

Examining the four lines which come after the END of the program will indicate that the purpose of the current test is to compare two alternative ways of finding the square of X, where X is a random number between 0 and 1. It is a convenient practice to put alternative versions of lines of program at the end of the program, rather as the spare wheel is tucked away in the back of a car, when using a microcomputer. Only two or three simple steps are needed to turn line 150 into line 50 and hence incorporate it into the program when required.

Our microcomputer could be made to record times in sixtieths of a second, and we then took the difference in time between a run of length 100 and a run of length 1100 as the time per 1000 performances of the loop formed by lines 30 to 60 of the program. With line 50 omitted the time was 10.1 seconds, while using lines 150 and 250 as versions of line 50 produced times of 58.2 seconds and 14.5 seconds respectively. So the time per calculation is 0.0481 seconds for line 150 and 0.0044 seconds for line 250. The calculation X↑2 takes eleven times as long as X*X because it works by taking the logarithm of X, doubling it, and then taking the antilogarithm; it must be borne in mind that functions such as LN, LOG, EXP, SIN, COS, etc., require a lot of calculations and so take a long time. There may be some versions of BASIC, however, which can interpret X↑2 as X*X and are accordingly fast.

Another example of a speed test which is of particular interest in statistics is the testing of alternative methods of calculating the geometric mean of a set of numbers. The lines needed for this test can be added to Program 3.5:

Program 3.5 (extension)

```
300 REM LINES 325 TO 365 BECOMES LINES 25 TO 65 FOR
    METHOD GM1
325 PG = 1
350 PG = PG*X
365 GM = PG↑(1/N)
400 REM LINES 425 TO 465 BECOMES LINES 25 TO 65 FOR
    METHOD GM2
425 GM = 1: P = 1/N
450 GM = GM*X↑P
465 REM
```

```
500 REM LINES 525 TO 565 BECOME LINES 25 TO 65 FOR
    METHOD GM3
525 SG = 0
550 SG = SG + LN(X)
565 GM = EXP(SG/N)
```

It is a good idea to insert a temporary line such as 355 PRINT PG↑(1/I) to make sure that the program is doing what it is required to do, this line being deleted before the timings are carried out. It will be found that Method GM1, although by far the fastest of the three methods, carries a high risk that PG will either collapse to zero (as in Program 3.1) or become too large (as in Program 3.2). Method GM2 takes the power of a number within the loop and so takes about 0.05 seconds, while Method GM3 only takes a logarithm within the loop and so is almost twice as fast. So Method GM3 is best.

A careless programmer might use the line 450 GM = GM*X↑(1/N) instead of using the variable P. This would add about 0.004 seconds to the calculation time for the loop. Everything possible should be done to minimize the calculation time within the loop if it is to be performed a large number of times.

3.3 Computing the mean and variance

In discussing some problems of accuracy in statistical calculations, it is convenient first to present a program which is developed from Program 2.8. Measures of location other than the arithmetic mean have been left out, and so the arithmetic mean is termed simply 'Mean' as is customary. The interactive statements INPUT have been replaced by READ . . . DATA statements; the reader should acquire facility in making the necessary changes since later programs in this text will be offered in one or other form but not in both.

Program 3.6

```
10 PRINT: PRINT: PRINT "CALCULATION OF MEAN AND
   VARIANCE"
20 PRINT: PRINT "Is the data strictly positive? ";: READ P$:
   PRINT P$
30 PRINT "Is the data collected by frequencies? ";: READ F$:
   PRINT F$
40 PRINT: IF F$="Y" THEN PRINT "Values with their
   frequencies:"
50 IF F$="N" THEN PRINT "Observations:"
60 PRINT "Input 1E+11 to end input."
70 PRINT: F=1
```

```
80 N = 0: S1 = 0: S2 = 0
90 READ X: IF X>1E+10 GOTO 130
100 PRINT X;: IF F$="Y" THEN READ F: PRINT F,
110 N = N+F: S1 = S1+F*X: S2 = S2+F*X*X
120 GOTO 90
130 PRINT: PRINT: PRINT N; "observations"
140 M = S1/N: PRINT: PRINT "Mean = ";M
150 V = (S2-M*S1)/(N-1): PRINT: PRINT "Variance =";V
160 SD = SQR(V): PRINT "Standard Deviation =";SD
170 PRINT "Coefficient of Variation ";: IF P$="Y" THEN
    PRINT "=";SD/M
180 IF P$="N" THEN PRINT "does not exist."
190 PRINT: INPUT "Do you want another set of data? (Y/N)
    ";R$
200 IF R$="Y" GOTO 10
240 DATA N,Y,0,144,1,91,2,32,3,11,4,2,1E+11
250 DATA Y,Y,10,2,12,3,14,4,16,5,17,6,19,8,20,10,21,12,1E+11
260 DATA Y,N,1231,1233,1234,1235,1237,1E+11
270 DATA Y,N,12342,12344,12345,12346,12348,1E+11
280 DATA Y,N,123453,123455,123456,123457,123459,1E+11
290 END
```

It is unnecessary to insert '(Y/N)' after a question if the user is accustomed to replying always with 'Y' or 'N', but in any case the answers to the questions at lines 20 and 30 are being read from the DATA. Line 190 provides a delay so that the output can be read from a screen, but would need to be adapted if the output is on a printer.

The first two sets of data may be recognized as Data 3 and Data 5 from the previous chapter. There are then three new sets of data:

Data 8

1231 1233 1234 1235 1237

Data 9

12342 12344 12345 12346 12348

Data 10

123453 123455 123456 123457 123459

Data 8 has mean 1234 and variance exactly 5. Each observation in Data 9 is 11111 more than the corresponding observation in Data 8, and so the mean is 12345 and the variance is again 5. Similarly, each observation in Data 10 is 111111 more than the corresponding observation in Data 9, and so the mean is 123456 and the variance is once again equal to 5.

However, many versions of BASIC will produce zero as the variance of Data 10. All versions of BASIC will eventually fail to find the variance if we keep adding further significant figures to the data while retaining the deviations from the mean at –3, –1, 0, 1, 3, respectively so that the true sum of squared deviations remains at 20 and the variance remains at 5.

One way to avoid this difficulty is to put all the data into an array, calculate the mean, and then calculate the deviations and sum the squared deviations so that for Data 10 it is never necessary to square a 6-digit number. There is however another method which does not require an array or inputting the data twice. It involves carrying a running value SS for the sum of squared deviations instead of carrying the value S2 for the sum of squared observations.

We make use of the fundamental relationship for any constant a:

$$\sum_{i=1}^{N}(x_i-\bar{x})^2 = \sum_{i=1}^{N}(x_i-a)^2 - N(\bar{x}-a)^2$$

Putting a equal to zero produces the relationship given in Section 2.7. It may also be noted that, since $(\bar{x}-a)^2$ cannot be negative, the relationship offers a proof that the sum of squared deviations about the 'centre' of the data is minimized by choosing $\bar{x}$ as the 'centre'.

For our present purpose, we apply the relationship when the Nth observation is inputted by letting a be the mean of the previous $(N-1)$ observations and SS be the sum of squared deviations of the same $(N-1)$ observations about that mean. To find the new value of SS, represented by the left-hand side of the relationship, we must add $(x-a)^2$ where x is the Nth observation and a is still the 'old' mean and then subtract $N(\bar{x}-a)^2$ where $\bar{x}$ is the 'new' mean. In the modification to Program 3.6 we use A and M for a and $\bar{x}$ respectively, and adapt the method to accept F observations of value X instead of only a single observation.

Program 3.6 (modification)

```
 80 N = 0: S1 = 0: M = 0: SS = 0
110 N = N+F: S1 = S1+F*X: A = M
115 M = S1/N: SS = SS+F*(X–A)*(X–A)—N*(M–A)*
    (M–A)
140 PRINT: PRINT "Mean = ";M
150 V = SS/(N–1): PRINT: PRINT "Variance = ";V
```

No special lines are needed for the first value of X to be inputted, other than putting M = 0 in line 80; the effect is to make M equal to X and SS equal to zero on the first performance of line 115, and this is correct. It will be found that the program loses accuracy slightly — such as 5.0000002 for the variance of Data 8 — because far more manipulation of non-integer quantities is performed.

Even the modified program is unable to cope with the following:

Data 11 1E + 15 3E + 15 4E + 15 5E + 15 7E + 15

Clearly the mean is 4E + 15 and the variance is 5E + 30. It is left as an exercise for the reader to find how a program can be made to cope with this data, provided that it can handle numbers up to 1E + 18. It will be necessary to divide all the observations by, say, 1E + 8 at some stage and to print out the variance as 5E + 14 MULTIPLIED BY 1E + 16.

3.4 Exact integer arithmetic

It has already been seen that the use of X↑2 to square a number is very slow. We may now examine its accuracy, in a program designed to test the dimensions of right-angled triangles.

Program 3.7

```
 10 PRINT: PRINT: INPUT “Input sides of triangle: ”;A,B,C
 20 IF A>B THEN X = A: A = B: B = X
 30 IF B>C THEN X = B: B = C: C = X
 40 PRINT: PRINT “The sides and their squares are:”
 50 A2 = A↑2: B2 = B↑2: C2 = C↑2: S = A2+B2
 60 PRINT: PRINT A;A2,B;B2,C;C2
 70 PRINT: PRINT “The sum of the first two squares is”;S:
    PRINT
 80 IF S=C2 THEN PRINT “This is a right-angled triangle.”
 90 IF S<>C2 THEN PRINT “This is NOT a right-angled
    triangle.”
100 GOTO 10
```

It will be found with many BASIC interpreters that hardly any right-angled triangle, even the simple 3, 4, 5 triangle, will be recognized by this program. Moreover, the more willing a program is to recognize right-angled triangles correctly the more likely it is to do so incorrectly.

The program is improved by avoiding the statements which involve the logarithmic function. Line 50 becomes:

```
50 A2 = A*A: B2 = B*B: C2 = C*C: S = A2+B2
```

Triangles to be tested on the alternative versions of the program are: 3,4,5: 5,12,13: 8,15,17: 16329,38980,42262: 36329,60721,70759. The first three triangles are right-angled; the other two are not, as may be seen immediately from the final digits of the numbers, but are liable to be treated as right-angled because of rounding errors.

Clearly there is need for a program which will carry out exact integer arithmetic. The following program, which is easily understood if carefully studied, will multiply together two numbers each of which may have any number of digits provided that the product has not more than 765 digits, and subject to restrictions on the length of strings.

Program 3.8

```
 10 DIM A(255),B(255),C(255)
 20 T = 1000: Z$ = "000"
 30 PRINT: PRINT "EXACT MULTIPLICATION OF
    INTEGERS"
 40 PRINT: INPUT "Input first number: ";A$
 50 LA = LEN(A$): NA = INT((LA-1)/3)
 60 FOR I = 0 TO NA-1
 70 A(I) = VAL(MID$(A$,LA-2-3*I,3))
 80 NEXT I
 90 A(NA) = VAL(LEFT$(A$,LA-3*NA))
100 PRINT: INPUT "Input second number: ";B$
110 LB = LEN(B$): NB = INT((LB-1)/3)
120 FOR I = 0 TO NB-1
130 B(I) = VAL(MID$(B$,LB-2-3*I,3))
140 NEXT I
150 B(NB) = VAL(LEFT$(B$,LB-3*NB))
160 NC = NA+NB+1
170 FOR I = 0 TO NC: C(I) = 0: NEXT I
180 FOR I = 0 TO NA
190 FOR J = 0 TO NB
200 K = I+J: X = A(I)*B(J)+C(K)
210 Y = INT(X/T): C(K) = X-T*Y
220 IF Y>0 THEN K = K+1: X = Y+C(K): GOTO 210
230 NEXT J
240 NEXT I
250 PRINT: PRINT "Product =";
260 IF C(NC) = 0 THEN NC = NC-1
270 PRINT C(NC)
```

```
280 FOR I = NC–1 TO 0 STEP –1
290 PRINT ",";RIGHT$(Z$+STR$(C(I)),3);
300 NEXT I
310 GOTO 30
320 END
```

It is relatively simple to extend this program so that the product is transferred from array C to array A and can then be multiplied by a further number. Addition and subtraction of large integers could also be introduced as an option, but division by exact integer arithmetic is much more difficult to achieve.

It may finally be noted that the program breaks down if either of the numbers inputted is less than 10 or if the product is less than 1000. The reader may or may not feel that it is worthwhile making the program 'foolproof' in this respect.

3.5 Ill-conditioned equations

It is important to recognize that small inaccuracies in the values inputted or in the results of intermediate calculations may lead to much larger inaccuracies in the final results. The problem of such 'ill-conditioning' arises very frequently in advanced statistical computing, such as in estimating the coefficients of regression equations. A simple example will illustrate the effect of ill-conditioning in solving sets of simultaneous linear equations.

Consider the pair of equations:

$$11.2\,x + 13.2\,y = 240$$
$$21.6\,x + 27.6\,y = 480$$

which could arise from a problem of the form 'How many units of each product should we produce per hour to keep all the machines fully occupied?'. Such practical problems often give equations in which one equation is almost an exact multiple of another.

The solution is $x = 12$, $y = 8$. However, if the coefficients are rounded to the nearest integer (perhaps using the argument that manufacturing times cannot be relied on more accurately than to the nearest minute) the solution becomes $x = 21.8$, $y = 0$. The enormous change in the solution has been effected by changes of less than 2 per cent in the coefficients.

A computer would not of course treat 11.2 as if it were 11, and so it would produce the correct solution to the original equations. But a good test of a computer using BASIC would be the equations:

$$8.099999\,x + 9\,y = 10$$
$$9\,x + 10\,y = 11.111112$$

The correct solution is $x = 0.8$, $y = 0.3911112$. A typical solution using BASIC is $x = 0.805$, $y = 0.387$, which is hardly satisfactory.

Where there are a large number of variables and of equations the problem of ill-conditioning is much more likely to arise. Small numerical rounding errors at early stages in the solution procedure can lead to much larger errors at later stages and to completely erroneous or nonsensical solutions. Good regression programs have to be designed to minimize the effects of ill-conditioning and to give warnings if the solutions are liable to be inaccurate.

PROBLEMS

(3.1) Run Programs 3.1 to 3.4 on your computer, and record for future reference the smallest and largest values accepted. Also find the smallest difference between X and Y such that X = Y in a conditional statement will not be treated as true. Find the smallest positive quantity such that SGN(X) is treated as 1 rather than 0. If the computer responds to PRINT π or PRINT PI, compare it for accuracy with a good pocket calculator. (Our calculator responded to the π button with 3.1415926, but when this 8-digit number was subtracted the remainder was 5.36 E−08 giving an 11-digit value for π.)

(3.2) Carry out the speed tests in Section 3.2 on your computer. Test the speeds of any other functions or procedures you may be interested in. Keep a notebook of (i) the 'translations' needed for your computer's BASIC, (ii) the limitations found in carrying out Problem 3.1, and (iii) any useful information on the relative speeds of different types of statement.

(3.3) Amalgamate Programs 3.5 and 3.7 to produce a program which tests both the speed and the accuracy of alternative methods of squaring random numbers or of numbers inputted by the user. (There is nothing wrong in asking the computer to square the same number 1000 times if this is necessary to obtain a good estimate of the time for a single calculation. Computers have unlimited patience, and never complain when required to do apparently useless repetitions; it would sometimes be better if they did complain.)

(3.4) As suggested at the end of Section 3.3, write a program that will calculate correctly the variance of Data 11. Try to incorporate this improvement into Program 3.6 as modified, to produce a program which will cope with Data 10, Data 11, and every other kind of awkward data.

(3.5) As suggested at the end of Section 3.4, try to extend Program 3.8 so that it will carry out repeated multiplication, addition, and subtraction operations. Also make it resistant to the inputting of integers less than 10. (A program is 'resistant' to an incorrect input if it responds by saying what is wrong and asking for new input. Better still would be to make the program handle all data correctly.)

(3.6) In later programs we shall need to use 1/SQR(2*π) as a constant. Find its value correct to 10 significant figures, using Program 3.8 and the fact that $\pi = 3.1415926536$. (Hint: the approximate value is 0.4. Multiply 40 by 40 and then multiply the product by 31415926536. The result begins with 5, so try again with 39 instead of 40. Keep on until you find the 10-digit integer which when squared and multiplied by 31415926536 produces a result as close as possible to 5 times a power of 10.)

(3.7) Write a program which will solve a pair of simultaneous linear equations. A simple program will deal with pairs of equations which have a unique solution; try to develop it to deal also with contradictory equations or equations with no unique solution.

(3.8) Test your computer on the ill-conditioned equations given in Section 3.5, using the program for Problem 3.7 if you have written it. Try the equations:

$$x + 2y = 3$$
$$2.0001\,x + 4.0001\,y = 6.0002$$

Treating the second equation as $(2 + \delta)x + (4 + \delta)y = 6 + 2\delta$, the solution will be $x = 1$, $y = 1$, whatever the value of δ. Keep decreasing the value of δ to see how the solutions become inaccurate when δ is very small, and eventually the program fails to find any solution.

Chapter 4

Elementary probability theory and simulation

4.1 Introduction to probability

It is not possible to give a full explanation of probability and statistics, even at an elementary level, in a text which is primarily concerned with BASIC programming in statistics. Instead, we shall aim to provide a precise but succinct account of the basic concepts and formulae with a minimum of illustrations. This will enable the informed reader to refresh his knowledge. The reader who is new to the subject will be provided with the means to test the soundness of his textbook in statistics, taking note of our comments in the Preface and Bibliography.

The question arises whether it is necessary to understand probability in order to apply statistical techniques. The answer is in the negative if one is interested only in descriptive statistics. It is also possible to carry out routine work in statistical inference, such as estimation and hypothesis testing, without any understanding of probability theory; but it is essential to have such understanding in order to interpret the results correctly.

4.2 Essential definitions

Whenever we discuss probability, it is with reference to some kind of experiment or investigation which has more than one possible outcome. Statisticians use the term *outcome* to mean a 'possible outcome'. The set of all outcomes is termed the *sample space* — though 'outcome set' or 'possibility set' might well have been better names.

The term *set* is used in its precise mathematical definition, as a collection of objects or abstractions which are the *elements* of the set. The elements are sometimes called *sample points.* The term *experiment* is used in its widest sense to cover any procedure which gives rise to a sample space comprising more than one outcome; so tossing a coin, throwing a six-sided die, or having a baby whose sex is not predetermined, are all 'experiments'. For convenience we use the word 'experiment' as shorthand for 'experiment or investigation'; an investigation might consist of selecting 20 male students at random and recording for each *specimen* the age in years, height in centimetres, weight in kilograms, shoe size, and number of teeth. Each piece of information recorded would be termed an *observation* whether obtained by observing, measuring, or questioning the specimen.

The elements of the sample space must be defined in such a way that the result of the experiment will always correspond to one and only one sample point. Most of the fundamental concepts of elementary probability can be illustrated by such simple experiments as tossing a coin or throwing a six-sided die, but later we shall consider experiments consisting of more than one 'trial' such as two or more tosses of a coin or two throws of a die. The sample space always consists of the outcomes of the *whole experiment,* however many trials it comprises.

In throwing a die (which we shall henceforth take to be always a six-sided die), the outcomes are the numbers on the faces which can appear uppermost after the throw. The label S is often given to the sample space. So for this experiment,

$$S = \{1,2,3,4,5,6\}$$

An *event* is a subset of the sample space. That is, it is a collection of outcomes, a single outcome, the empty set, or the whole sample space. The empty set is denoted by ϕ. For the experiment under consideration there are 64 different events, including ϕ and S; some examples are:

$$A = \text{at least } 3 = \{3,4,5,6\}$$
$$B = \text{less than } 5 = \{1,2,3,4\}$$
$$C = \text{an odd number} = \{1,3,5\}$$
$$D = \text{a six} = \{6\}$$
$$E = \text{an even number} = \{2,4,6\}$$

The *intersection* of two sets is the set of elements which belong to both sets. Two events whose intersection is the empty set are termed *mutually exclusive events.* Thus B and D are mutually exclusive since

$$B \cap D = \phi$$

where $\cap$ is the symbol for the intersection.

The *union* of two sets is the set of elements which belong to either set or to both sets. The symbol for the union is $\cup$, and so

$$B \cup C = \{1,2,3,4,5\}$$
$$B \cup D = \{1,2,3,4,6\}$$

If the union of two events constitutes the whole sample space they are said to be *exhaustive.* A and B are exhaustive since

$$A \cup B = S$$

Two events which are both mutually exclusive and exhaustive are termed *complementary events.* It can be seen that C and E are complementary. It follows that every outcome which is not in C must be in E, and vice versa, and so E can in this example be termed 'not C'. Each event has a unique complementary event.

It must be emphasized that we can only define events as 'mutually exclusive', 'exhaustive', 'complementary', or 'in none of these relationships' if the events are subsets of *the same sample space.* It will also be noted that the concept of probability does not enter into these definitions.

A *probability* is a number which is associated with an event in accordance with certain rules. The formal definition of probability is in terms of its axioms, but a satisfactory working definition is in terms of *limiting relative frequency.* Let A be an event, and let $N(A)$ be the number of times A occurs in N performances of the experiment. Then $N(A)$ divided by N is the *relative frequency* of A. The observed relative frequency will differ for different sequences of N performances of the experiment, but we postulate that for large N the relative frequency tends to a specific value and this limiting relative frequency is termed the *probability of* A and denoted by $P(A)$.

The 'limit' to which the relative frequency tends cannot be formally defined as a mathematician defines a limit. The postulate that the limit exists is based on belief in the consistency of nature; without such a belief it would be impossible to generalize scientific laws.

4.3 The laws of probability

If two events are mutually exclusive then the relative frequency of their union must be the sum of their separate relative frequencies. Taking the relative frequencies to their limits produces the *Law of Addition for mutually exclusive events*:

$$P(A \cup B) = P(A) + P(B) \text{ given that } A \cap B = \phi$$

The formal 'axiomatic' definition of probability takes this law as one of its axioms, the other two being

$$P(A) \geqslant 0 \text{ for any event } A$$
$$P(S) = 1$$

Both these results are expected when using the limiting relative frequency definition, and so there is no conflict between this informal definition and the formal axioms.

Sometimes we are able to assume, or wish to hypothesize, that all the outcomes are equally likely. Before making such an assumption we must be agreed as to which sample points form the fundamental sample space; sometimes an outcome as defined by one experimenter can be decomposed into two or more outcomes in the view of another experimenter. If there are n sample points in the fundamental sample space and all are equally likely then each must have probability $1/n$, and the probability of any event comprising m sample points must be m/n. These results follow from the Law of Addition and the fact that $P(S) = 1$.

If we say that an 'unbiased' coin is tossed or an 'unbiased' die is thrown, this is intended to indicate that all the sample points are equally likely. If an unbiased die is thrown the probability of an event comprising m sample points is $m/6$, and so the probabilities of the five events previously defined are:

$$P(A) = 2/3 \quad P(B) = 2/3 \quad P(C) = 1/2 \quad P(D) = 1/6 \quad P(E) = 1/2$$

Two events are said to be *independent* if the probability of their intersection is the product of their separate probabilities. This is the *Law of Multiplication for independent events*:

$$P(A \cap B) = P(A)\,P(B) \text{ if and only if } A \text{ and } B \text{ are independent.}$$

Since the law constitutes the definition of independent events, it is a tautology; it cannot be proved, and does not need to be proved.

In the example of the unbiased die, $P(B \cap C) = P(\{1,3\}) = 1/3 = P(B)\,P(C)$ and so B and C are independent. It can similarly be seen that B and E are independent. $P(A)\,P(B)$ in that example works out as 4/9, and since this is not a multiple of 1/6 it can be concluded that A and B are not independent without even examining their intersection.

The *conditional probability of B given A* is defined as

$$P(B|A) = \frac{P(A \cap B)}{P(A)}$$

It follows that:

$$P(A \cap B) = P(A)\,P(B|A).$$

Hence if we know that $P(B|A) = P(B)$ then A and B are independent. Similarly, by interchanging A and B throughout, we find that $P(A|B) = P(A)$ implies that A and B are independent and this in turn implies that $P(B|A) = P(B)$. We do not define independent events by the property $P(B|A) = P(B)$ since $\mathrm{P}(B|A)$ does not exist if A has zero probability. Under the correct definition it is clear that if $P(A) = 0$ then A is independent of every other event.

The final law is almost self-evident. The *Law of Complementarity* states that:

$$P(A) = 1 - P(A')$$

where A' is the complementary event of A. Like the other laws, it should be regarded as a tool rather than as a constraint; it is often easier to find the probability of a complicated event A by considering the probability of A' and subtracting it from 1 than by analysing the sample points contained in A.

An event is *certain* if it constitutes the whole sample space, and *impossible* if it is the empty set. $P(A) = 0$ does not imply that A is impossible; it is not impossible for a man to be exactly 180 cm tall but the probability of this sample point is zero. So it is not correct to define mutually exclusive events by $P(A \cap B) = 0$ or exhaustive events by $P(A \cup B) = 1$.

4.4 Multi-trial experiments

All the laws of probability have been defined and illustrated while considering only a 'single-trial' experiment. If an experiment consists of a number of actions, such as two or more tosses of a coin or two or more throws of a die, it is convenient to refer to the separate actions as 'trials'. We can often find the probabilities of the outcomes of the whole experiment by considering the sample spaces for the separate trials.

We define two trials as *independent trials* if any event which is defined solely by reference to what happens on one trial is independent of any event which is defined solely by reference to what happens on the other trial. In brief, they are independent trials if they 'do not affect each other' — a phrase which can *not* be used to define independent events.

In the experiment of throwing a die twice, there are 36 sample points in the sample space. If we can show (or are prepared to assume) that the probability of the event '1 on the first throw and 1 on the second throw' is the product of the probabilities of the events '1 on the first throw' and '1 on the second throw', and similarly for all the other 35 pairs of values, then the trials are independent. This is true irrespective of whether or not the die is unbiased.

If the physical conditions of the experiment are such that the throws 'do not affect each' other in a practical sense then they will be independent trials. When statisticians apply the corpus of statistical inference of which the foundations are the Laws of Addition and

Multiplication, they do not indulge in detailed investigation of the probabilities of intersections of events; they try to ensure by the physical design of experiments that trials which are assumed to be independent cannot in fact affect each other.

When an unbiased die is thrown twice and the throws do not affect each other, the 36 sample points are all equally likely. Consider three out of the millions of events which could be defined on this sample space

A = an even number on the first throw
B = the sum of the numbers is 7
C = the sum of the numbers is 8

It will be found that A and B are independent but A and C are not independent. B and C are of course mutually exclusive. The ability to recognize these facts is a good test of one's understanding of elementary probability.

We shall however be concerned later with experiments in which two or more specimens are selected without replacement from a finite population. If two cards are selected without replacement from a 52-card pack, there are 52×51 sample points in the sample space. If r cards are selected at random without replacement from a 52-card pack, there are $52 \times 51 \times \ldots \times (53-r)$ sample points in the sample space and the inclusion of the phrase 'at random' indicates that they are all equally likely. Each sample point is a *permutation*. Any specified set of r cards can be selected in $r!$ different ways (where $r! = 1 \times 2 \times 3 \times \ldots \times r$) which together form a *combination*. Since all combinations contain the same number of sample points and all sample points are equally likely, it follows that all combinations are equally likely. The reader should refresh his understanding of permutations and combinations and of the notations ${}_nP_r$ and ${}_nC_r$, and should note that problems in probability which involve combinations rather than permutations are often simplest dealt with by considering the numbers of equally-likely combinations without reference to the fundamental sample space in which each permutation is a sample point.

The fundamental definitions of probability and laws of probability have been dealt with sufficiently carefully for the reader to test out his own knowledge or the soundness of the textbook which will henceforth be assumed to be at his left hand. We shall add one further result which is sometimes of use and is known as the 'general additive rule'.

$$P(A) + P(B) = P(A \cup B) + P(A \cap B)$$

This general additive rule can be re-arranged as necessary to find the fourth probability when the other three are known. It is also useful if two probabilities are known together with a relevant relationship:

If A and B are independent: $P(A \cup B) = P(A) + P(B) - P(A)\,P(B)$
If A and B are exhaustive: $P(A \cap B) = P(A) + P(B) - 1$.

We conclude the section with a short and self-explanatory program.

Program 4.1

```
10 PRINT: PRINT: PRINT" TWO-TRIAL SAMPLE SPACE"
20 PRINT: INPUT "Input number of outcomes for the first trial
   ";N1
30 DIM A1$(N1),P1(N1): T = 0
40 PRINT: PRINT "Input description of sample point, then
   probability."
50 FOR I = 1 TO N1
60 PRINT: PRINT "No.";I;":";: INPUT A1$(I),P: T = T+P:
   P1(I) = P
70 NEXT I
80 IF ABS(1-T)>0.01 THEN PRINT: PRINT "WARNING:
   Total probability =";T
90 PRINT: INPUT "Input number of outcomes for the second
   trial ";N2
100 DIM A2$(N2), P2(N2): T = 0
110 PRINT: PRINT "Input description of sample point, then
    probability."
120 FOR I = 1 TO N2
130 PRINT: PRINT "No.";I;": ";: INPUT A2$(I),P: T = T+P:
    P2(I) = P
140 NEXT I
150 IF ABS(1-T)>0.01 THEN PRINT: PRINT "WARNING:
    Total probability =";T
160 PRINT: PRINT
170 PRINT "The listed probabilities assume that the trials are
    independent."
180 PRINT
190 FOR I = 1 TO N1
200 FOR J = 1 TO N2
210 PRINT A1$(I),A2$(J),P1(I)*P2(J)
220 NEXT J
230 NEXT I
240 END
```

It should be noted that there is no reason why both trials in a two-trial experiment should have the same individual sample space, and the above program has been designed to deal with dissimilar trials.

4.5 Introduction to simulation

It is almost impossible to overstate the importance of statistical simulation in all branches of industry and commerce with the advent of cheap and fast microcomputers and microprocessors. To take only one example, on-line microprocessors can record all the production variables of a manufacturing process when it is 'in control' and when it is going 'out of control'. Analysis of this data will permit all the relevant statistical properties to be identified or estimated. Large-scale simulations of the process both when it is in control and when it is out of control or going out of control can then be run on a computer, and a variety of quality-control rules evaluated to see which rules are most successful in 'catching' the machine quickly when it is going out of control without producing a large number of 'false alarms'.

We could enumerate other applications in project-planning, investment-evaluation, sales-forecasting, etc. The biggest limitation on the application of microcomputers employing simulation techniques is the lack of knowledge and imagination on the part of the managers who are responsible for making the decisions. Another limitation is the lack of management scientists who are skilled at simulation and aware of the pitfalls. In this text there is of course room only for a consideration of one or two basic ideas.

To introduce the idea of computer simulation, we offer a very simple program to simulate the throw of a die.

Program 4.2

```
 10 DIM N(5)
 20 PRINT: PRINT: PRINT "SIMULATION OF THE VALUE
    THROWN ON A DIE"
 30 PRINT: INPUT "Input length of run";N
 40 FOR I = 1 TO N
 50 R = INT(6*RND(1))
 60 N(R) = N(R)+1
 70 NEXT I
 80 PRINT
 90 FOR I = 0 TO 5: PRINT I+1,N(I),N(I)/N: NEXT I
100 END
```

Putting N equal to 600, so that on average each value should occur

100 times, means a running time of about 15 seconds on a typical microcomputer.

A common error in simulation studies is to hopelessly underestimate the length of run required to get useful results. Many operational researchers think that a run of 1000 simulated performances of an experiment is very large, and never dream of a run longer than about 5000.

The true probability for the value 1 on a throw of a die is 0.167. To have a fairly good chance of getting within 0.01 of this value the simulation run needs to be of length 6000. To have a similar chance of getting within 0.001 of the true value the run needs to be of length 600 000! The secret of good simulation is to ensure that the random-number generator produces high-quality random numbers, to design the program very carefully to maximize the speed of the simulation loop and to extract the maximum amount of information (for instance, to test all conceivable quality-control rules simultaneously on the same simulation run), and when everything has been tested as carefully as a moon rocket is tested before launching, to carry out a massive simulation run overnight or over the weekend.

To simulate the sum of the numbers thrown on two dice, or of the numbers obtained when throwing a die twice, only four lines of Program 4.2 need to be changed:

```
10 DIM N(10)
20 PRINT: PRINT: PRINT "SUM OF NUMBERS THROWN
   ON TWO DICE"
50 R = INT(6*RND(1))+INT(6*RND(1))
90 FOR I = 0 TO 10: PRINT I+2,N(I),N(I)/N: NEXT I
```

Line 50 simulates the sum of two independent random integers each of which is equally likely to take any value from 0 to 5. The sum is corrected for dice numbered from 1 to 6 at line 90, which is outside the simulation loop.

As discussed in the previous section, there are 36 equally-likely sample points in the sample space. The 'sum of numbers thrown' produces a set of 11 events:

Sum of numbers thrown	*Number of sample points*
2	1
3	2
4	3
5	4
6	5
7	6
8	5

Sum of numbers thrown	*Number of sample points*
9	4
10	3
11	2
12	1

The probabilities of the events are obtained by dividing the respective numbers of sample points by 36. If the relative frequencies in the simulation run fail to converge on these probabilities, it must be because the random-number routine is defective. It is important to know whether this is the case, since we may want to use simulation experiments in cases where the true probabilities are not known.

A run of length 3600 on a popular microcomputer aroused strong suspicions that the random-number generation was defective, and this was confirmed by a run of 18 000:

Sum of numbers	*Observed frequency*
2	488
3	1021
4	1552
5	2091
6	2629
7	2346
8	2617
9	2095
10	1603
11	1023
12	535

The low number of occurrences of 7 first gave rise to suspicion. It can be shown that in 18 000 trials an event which has probability 1/6 has an observed frequency with mean 3000 and standard deviation 50. The probability of getting a result which is more than 6 standard deviations below the mean is about 10^{-9}; we have a result which is more than 13 standard deviations below the mean!

An alternative method of simulating the result of the two-dice experiment is to find the appropriate formula to obtain a random integer with the required probabilities for the various 'sums of values' using only a single random number to generate it. The way to find the formula is beyond the scope of this text, but it is not difficult to check that the formula is correct. The formula generates integers between 2 and 12 rather than between 0 and 10, so we offer a complete new program:

Program 4.3

```
 10 DIM N(12)
 20 PRINT: PRINT: PRINT "SUM OF VALUES THROWN ON
    TWO DICE"
 30 PRINT: INPUT "Input length of run";N
 40 FOR I = 1 TO N
 50 X = RND(1): IF X<0.5 THEN R = INT((3 + SQR(1 +
    288*X))/2)
 55 IF X> =0.5 THEN R = INT((27–SQR(289–288*X))/2)
 60 N(R) = N(R)+1
 70 NEXT I
 80 PRINT
 90 FOR I = 2 TO 12: PRINT I,N(I),N(I)/N: NEXT I
100 END
```

It will be seen that if X is less than 7/12 then R will be less than 8 while if X is less than 5/12 then R will be less than 7, and so the probability that R is 7 will be the probability that X is between 5/12 and 7/12 which will be exactly 1/6 if RND(1) is unbiased. All other values may be similarly checked.

The results for a run of 18 000 were

Sum of numbers	*Observed frequency*
2	502
3	1011
4	1527
5	1993
6	2494
7	3087
8	2511
9	1939
10	1496
11	946
12	494

These give a satisfactory fit to the theoretical probabilities. The two methods of generating the random integers have in fact shown convincingly that the 'pseudo-random-number generating procedure' on this microcomputer is adequate when producing single random numbers, for which the only requirement is that they are equally likely to lie anywhere between 0 and 1, but seriously defective in respect of the requirement that successive random numbers should be independent.

4.6 Producing high quality pseudo-random numbers

For serious simulation it is essential to be able to generate pseudo-random numbers which are sound, and desirable to be able to 'copy' the sequence if necessary. There are satisfactory methods based on the formula:

$$R_{n+1} = M*R_n - D*\text{INT}(M*R_n/D)$$

which will produce random integers between 1 and $D-1$ provided that M and D are suitable integers. Taking for illustrative purposes the integers 10 and 97 (far too small for serious use), and setting $R_0 = 1$, we obtain the sequence:

$$R_1 = 10*1 - 97*\text{INT}(10*1/97) = 10 - 0 = 10$$
$$R_2 = 10*10 - 97*\text{INT}(10*10/97) = 100 - 97 = 3$$
$$R_3 = 10*3 - 97*\text{INT}(10*3/97) = 30 - 0 = 30$$

The sequence continues 9, 90, 27, 76, 81, 34, 49, etc. These values of M and D are useful for setting-up and testing a pseudo-random-number generating program, but must then be replaced by a better pair of values. It is known that an excellent pair are $M = 8192$, $D = 67101323$.

A difficulty remains, however. The sequence will go seriously wrong if the integer arithmetic is not exact, and this requires the multiplication of 8192 by 67 101 322 which exceeds the capacity of most computers using BASIC as we saw in Section 3.1. The solution is to split up the large integers into two or more separate integers, such as one to represent multiples of 10 000 and the other to represent the remainder after dividing by 10 000. The arithmetic is carried out rather on the lines of that in Program 3.8, but requires also a routine for division.

Producing such a procedure requires some ingenuity. The program now to be presented succeeds by using multiples not of 10 000 but of M. Its speedy method of dividing by D depends on the fact that D is greater than $M(M-1)$ but less than M^2, which is true for both the pairs of values we have considered. It also requires that the BASIC will perform exact integer arithmetic up to M^2, which was true for our microcomputer but may not be the case for all BASIC users. It is convenient to test the program using $M = 10$ and $D = 97$, and the reader may find it interesting to use these values in a pencil-and-paper run of the program to see how it works. Note that the sequence produced consists of random integers, but that for practical use we shall want to divide the random integer by D to obtain a random number between 0 and 1.

Program 4.4

```
10 M = 8192: D = 67101323
20 B = M*M–D
30 PRINT: PRINT: PRINT "GENERATING HIGH QUALITY
   RANDOM NUMBERS"
40 PRINT: INPUT "Input length of run ";N
50 PRINT: PRINT "Input seeding integer (1 to";D–1;") ":
   INPUT S
60 R = S
70 FOR I = 1 TO N
80 C = INT (R/M)
90 R  = B*C + M* (R–M*C)
100 IF R>D THEN R = R–D
110 PRINT R,R/D
120 NEXT I
130 PRINT: PRINT "Seeding integer =";S: PRINT "Terminal
    integer =";R
140 GOTO 30
150 END
```

Of course, line 110 is likely to be replaced in practical use by some simulation operation making use of the random number R/D or some multiple thereof. Line 130 offers a reminder of the seeding integer and a record of the terminal integer. If $S = 1$ and $N = 10\,000$ then the sequence will always be the same, and this can be useful when trying to compare the success of different servicing systems when faced with the same sequence of random arrivals. The terminal integer would then be 42 995 302, and the experimenter would know that by using this as his next seeding integer he could produce a new sequence of 10 000 random numbers; moreover, combining the results of the two runs would exactly reproduce the effect of a single run of length 20 000.

While repeating our assertion that the potentialities of statistical simulation are almost impossible to overstate, we must re-emphasise the importance of designing the simulation experiments carefully and using high-quality random numbers, preferably in the form of a reproducible sequence. A little learning is a dangerous thing, in statistics perhaps more than in any other field. There must be more than one research student like the one we discovered who thought he could dispense with the large integers 8192 and 67 101 323 and work with fractions. In effect his method was as follows:

Program 4.5

```
 10 M = 8192: D = 67101323
 20 PRINT: PRINT: PRINT "OPTIMISTIC RANDOM
    NUMBERS"
 30 PRINT: INPUT "Input length of run ";N
 40 PRINT: PRINT "Input seeding integer (1 to";D-1;")":
    INPUT S
 50 X = S/D
 60 FOR I = 1 TO N
 70 X = M*X: X = X-INT(X)
 80 PRINT X
 90 NEXT I
100 END
```

If computers could calculate fractions exactly, this program would achieve the same results as Program 4.4. Since computer arithmetic is not exact, we pointed out that the method is bound to collapse eventually into a sequence of zeroes. But how quickly does it collapse? You may be as surprised as we were when it was put to the test.

4.7 Simulation of the two-armed bandit problem

It may be useful to apply simulation to a problem which is of great practical importance, has not yet been solved, and requires only a short program. As is often the case in probability, the model is defined by reference to gambling.

Consider a man who has n coins which he intends to insert into one or other of two 'one-armed bandit' gambling machines, with the object of maximizing the total number of 'wins'. It may be that one machine offers a higher probability of success than the other, but he can only assess the probabilities by trial-and-error. He cannot simply carry out $n/2$ trials on each machine, since the object is not the academic one of finding out the respective probabilities but the practical one of maximizing the number of successes.

This problem originated in considering trials of alternative medical treatments, but is equally applicable to trials of alternative methods of erecting buildings or bridges. We wish to find the best decision procedure, where the decision at each stage is based on the current records of success for the two treatments, so as to minimize the mean number of patients who will die or of bridges which will fall down as a result of applying the less successful treatment.

It is customary to define as optimal the set of rules which minimizes the maximum average loss. The average loss is the average number of occasions (averaged over an infinite number of repetitions of the whole experiment) on which the inferior treatment will be applied multiplied by the difference between the two probabilities of success. To find the maximum average loss, all possible pairs of probabilities come into consideration. A maximum must exist, since if the probabilities are nearly equal the mean loss per trial will be small, while if they are far apart it will require (on average) very few trials before it becomes relatively certain which treatment should be favoured.

Program 4.6

```
 10 DIM N(2),P(2),R(2)
 20 PRINT: PRINT: PRINT "SIMULATION OF TWO-ARMED
    BANDIT PROBLEM"
 30 PRINT: INPUT "Input P1,P2 ";P(1),P(2)
 40 PRINT: INPUT "Input length of run ";N
 50 FOR I = 1 TO 2
 60 N(I) = 1: R(I) = 0: IF P(I)>RND(1) THEN R(I) = 1
 70 NEXT I
 80 FOR I = 3 TO N
 90 J = 1: K = R(1)-N(1)*R(2)/N(2): T = N(1)+R(2)
100 IF K<0 THEN K = K*N(2)/N(1): T = N(2)+R(1)
110 P = (1+K*ABS(K)/(K*K+T))/2: IF P<RND(1) THEN
    J = 2
115 PRINT I,K; TAB(23);P;TAB(36);J
120 N(J) = N(J)+1: IF P(J) > RND(1) THEN R(J) = R(J)+1
130 NEXT I
140 PRINT: PRINT "Summary:"
150 FOR J = 1 TO 2: PRINT P(J),R(J);" in";N(J);
    TAB(28);R(J)/N(J): NEXT J
160 PRINT: PRINT "Length of run =";N
170 PRINT: PRINT "Probabilities of selecting the methods:
    ";P;1-P
180 J = 1: IF P(2) > P(1) THEN J = 2
190 PRINT: PRINT "Estimated Loss ="; N*P(J)-R(1)-R(2)
200 PRINT "or"; (N-N(J))*ABS(P(1)-P(2))
210 GOTO 20
```

In this program, lines 50–70 initiate the simulation with the not-unreasonable rule that each treatment must be tried once. In lines 90–100, K is a measure of the relative success of treatment 1 to treatment 2 and so is negative if treatment 2 has the better record of proportionate successes. The exact form of K, and of T,is designed so that in line 110 we can produce a value P for the probability that treatment 1 ought to be selected on the next trial. No expertise has gone into the development of these formulae; they are simply an amateurish attempt to find a formula which makes P tend to 1 if treatment 1 has proved consistently better than treatment 2, and tend to 0 if treatment 2 has proved consistently better. Note however certain essential properties of the 'rule' to be applied; one must not abandon one treatment completely merely because it starts off badly, and on the other hand if P(1) is greater than P(2) we want P to tend to 1 in the long run and not merely to tend to P(1).

When a simulation program is being developed it is always best to include lines such as line 115, to print out details of what is happening. The simulation run will consist of only 3 or 4 steps while the output is studied to ensure that the program is working correctly. As the line numbering suggests, line 115 is to be discarded when a long run is to be undertaken.

It is not obvious how one should measure the loss in a particular simulation run, and so alternative formulae are used at lines 190 and 200. In the long run the average loss as obtained by either of these formulae will tend to the true average loss.

Some suggestions are made in the Problems for developing the program so that it automatically works out the average loss and converges on the maximum average loss for a particular set of decision rules, and for possible variations of the rules. It is not impossible, though unlikely, that one of our readers will find a better decision procedure than has so far been found by the experts. A cheap old microcomputer with only 1 K of memory can be extremely useful in this kind of simulation, averaging and maximizing large numbers of possibilities employing large numbers of simulation runs while switched on for day-and-night operation but temporarily disconnected from its visual display unit.

For serious research it would be essential to use high-quality random numbers. The trials must be regarded as independent; that is, the probability of a success on any trial depends on the type of trial but not on any previous successes or failures.

PROBLEMS

(4.1) Examine all the problems on elementary probability in any textbook on probability and statistics. Do they always make it clear whether trials are independent, whether dice are unbiased, whether sampling is with or without replacement, etc., if such assumptions are necessary in order to produce solutions? If not made explicit, are the assumptions reasonable in the situations described?

(4.2) A and B are two events defined on the same sample space. Show that if they are both mutually exclusive and independent then at least one of them must have zero probability. Show that if they are both exhaustive and independent then at least one of them must have probability 1.

(4.3) In the experiment of throwing an unbiased die twice, assuming that the throws are independent, for what values of r is the event of throwing a total of r independent of the event of obtaining an even number on the first throw? Is the event of throwing two even numbers independent of the event of throwing two equal numbers?

(4.4) An unbiased die is thrown ten times. Given that at least one 6 is thrown, what is the probability that two or more 6s are thrown? State any assumptions which you make, and specify exactly which laws of probability you are using at each stage.

(4.5) The coefficients of the quadratic equation $Ax^2 + Bx + C = 0$ are determined by throwing an unbiased die three times; A = first value, B = second value, C = third value. What is the probability that the equation will have (i) real roots, (ii) rational roots?

(4.6) In Program 4.1, lines 90–150 are very similar to lines 20-80. Formulate a subroutine to avoid the duplication of statements and so shorten the program. Develop the program so that it will cope with more than two trials.

(4.7) Develop Program 4.1 so that it will print out a vertical bar chart for probabilities, as Program 2.6 (extension) in Section 2.5 does for discrete observations.

(4.8) Run Program 4.2 in the original and modified versions and also Program 4.3 on your computer, with run lengths of at least 3600. Do the observed frequencies seem reasonable? Keep a record of them for goodness-of-fit tests in Chapter 11.

(4.9) Incorporate into Program 4.2 as modified the high-quality pseudo-random-number generator from Program 4.4. Confirm that it produces reasonable observed frequencies, with a run length of at least 3600.

(4.10) Run Program 4.5 on your computer to see how quickly the sequence of 'random numbers' becomes a sequence of zeroes. (This time you will not need a run length of 3600.)

(4.11) Develop Program 4.6 so that it incorporates the high-quality pseudo-random-number generator, carries out several runs for each pair of values of $P1$ and $P2$ in multiples of 0.1, considers multiples of 0.01 in the region where the maximum average loss has been found, and hence determines the maximum average loss fairly precisely. (It is sufficient to consider values of $P2$ less than or equal to $P1$, since interchanging *P*1 and *P*2 should not affect the result.)

(4.12) Experiment with alternative decision procedures in Program 4.6 (as developed in Problem 4.11) to try to reduce the maximum average loss. Consider (i) replacing *T* at line 110 by a fraction or multiple of *T*; (ii) making more radical changes in the formulae for *K*, *T*, and hence *P* at lines 90–110; (iii) applying each treatment two or three times at lines 50–70 instead of only once. A completely different decision procedure which could be investigated is to apply each treatment an equal number of times until one has achieved *D* successes more than the other, whereupon the more successful treatment is permanently adopted; it is suggested that *D* should be about 0.3*SQR(*N*) for good results, but experimentation would suggest a more exact formula. (It is believed that the minimum attainable value for the maximum average loss in *N* trials is about 0.32*SQR(*N*).)

Chapter 5

Further descriptive statistics

5.1 More measures of location and quantiles

From now on, all programs in this text will be complete programs which although originally developed on a microcomputer have been run (with essential minor modifications) on a mainframe computer. Formal 'Program Notes' will be offered for each program. Throughout the text we have deliberately avoided 'consistency', preferring to offer examples of input by INPUT, GET, or READ statements and output to printer or to screen. By comparing the programs, the reader will learn to adapt them to suit the computers available.

We again remind the reader that our purpose, apart from offering and explaining programs in BASIC, is to summarize the principal definitions and formulae in statistics but not to teach the subject for the beginner. Occasionally we offer explanations to supplement or correct those commonly found in textbooks in statistics, where these are relevant to our programs.

The first program in this chapter involves no new theory or definitions. It amalgamates Program 2.7, as adapted to calculate the mean, with Program 2.9. It also calculates the median and the mode.

Program 5.1

```
 90 DIM N(255)
100 PRINT: PRINT: PRINT "    STATISTICS"
110 PRINT: PRINT
120 INPUT "Input number of observations ";N
130 INPUT "Input minimum and maximum values ";MN,MX
140 PRINT: R = MX-MN
150 FOR I = 1 TO N
160 PRINT "Input observation No. ";I;": ";
170 INPUT X: N(X-MN) = N(X-MN)+1
180 NEXT I
190 N = 0: S1 = 0: FH = 0
200 FOR I = 0 TO R
210 F = N(I): N = N+F: IF F>FH THEN FH = F
220 IF F>0 THEN X = MN+I: PRINT X,F,N: S1 = S1+X*F
230 NEXT I
240 PRINT: PRINT N;" observations"
250 M = S1/N: PRINT: PRINT "Mean   = ";M
260 H1 = (N+1)/2: H = INT(H1): I = 0: T = 0
```

```
270 T = T+N(I): IF T<H THEN I = I+1: GOTO 270
280 D = I: IF H=H1 OR T>H GOTO 320
290 H = H+1: I = I+1
300 T = T+N(I): IF T<H THEN I = I+1: GOTO 300
310 D = (D+I)/2
320 PRINT "Median = ";MN+D
330 PRINT "Mode   = ";
340 M = M-MN: T1 = 0: T2 = 0
350 FOR I = 0 TO R
360 F = N(I): IF F = FH THEN PRINT ;MN+I;"  ";
370 IF F>0 THEN T1 = T1+F*ABS(I-M): T2 = T2+F*ABS(I-D)
    : N(I) = 0
380 NEXT I
390 PRINT: PRINT: PRINT "Mean deviation about mean = ";T1/N
400 PRINT "Mean deviation about median = ";T2/N
410 GOTO 100
420 END
```

Program notes

(1) Line 90 replaces the declaration DIM N(R) which appeared in Program 2.7 after R had been calculated. The mainframe computer would not accept a declaration in terms of a variable, but in any case the explicit declaration at line 90 is needed once we decided at line 410 to put the program into the 'come again' form.

(2) Note that there is a severe restriction on the type of data which the program will handle. It must be in single observations, and the observations must be integers lying within a known range which does not exceed 255. As long as there are no values smaller than MN or larger than MX as inputted at line 130, and (MX−MN) does not exceed 255, it does not matter if the actual minimum value is greater than MN or the actual maximum value is smaller than MX. All these requirements were also true of Program 2.7.

(3) Lines 120–250 are very similar to Program 2.7 as modified to calculate the Mean, but they also find the highest frequency FH.

(4) Lines 260–320 find the median. Most of the difficulties arise only when N is an even number; lines 290–310 are never needed when N is odd, since H is then equal to H1.

(5) The mean and median both having been found, lines 340–380 find the sums T1 and T2 of the absolute deviations from the mean and from the median in the same way as in Program 2.9. M and D are of course equal to (mean − MN) and (median − MN) respectively, and I is equal to (X−MN), and so the deviations are (I−M) and (I−D) respectively.

(6) With line 330 as introduction, lines 340–380 also print out (at line 360) all values whose frequency F is equal to the highest frequency FH. This is necessary if the data is bimodal or multi-modal, i.e. if two or more values have jointly-equal highest frequencies, but rather unsatisfactory if FH = 1.

(7) Lines 390–400 print out the two forms of the mean deviation, as in Program 2.9. As was pointed out in Section 2.7, the mean deviation about the median is the preferable statistic but the mean deviation about the mean is more often asked for.

(8) Data 2 and Data 3 from Section 2.5 can be fed directly into Program 5.1 without using Program 2.6 to collect it by frequencies. The extension to Program 2.6, printing out a vertical bar chart, could be added to Program 5.1.

Program 5.1 produced the sample median, which is the simplest example of a *quantile.* A quantile, sometimes termed a 'fractile', is a value such that a prescribed proportion of the observations are less than or equal to it. For the median the proportion is one-half.

The *quartiles* are the values Q_1 and Q_3 such that one-quarter of the observations are less than or equal to Q_1 and three-quarters of the observations are less than or equal to Q_3. Clearly we could define Q_2 in a similar manner, and this would then be the median.

Just as the median among 19 observations is the 10th value when the observations are in ascending order, so Q_1 and Q_3 are then respectively the 5th and the 15th values. It should be noted that the 15th value from the bottom is the 5th value from the top when there are 19 values.

Statisticians rarely try to identify the quartiles when N is so small that it is likely to matter whether Q_1 is taken as the next value below or the next value above the position $(N + 1)/4$ if that is not an exact integer. It would however be correct if N is 20 to interpret $(N + 1)/4$ to mean that Q_1 should be three-quarters of the 5th value plus one-quarter of the 6th value when the observations are in ascending order. Taking this as clue, it is left to the reader to develop Program 5.1 so that it will determine and print out the quartiles; an extension of the ideas applied in lines 260–310 is required.

Often Q_1 is referred to as the *lower quartile* and Q_3 as the *upper quartile.* The difference between them is termed the *interquartile range,* and $(Q_3-Q_1)/2$ is termed the *semi-interquartile range* (SIQR).

It is regrettable that the 'interquartile range' is now explicitly included in some examination syllabuses in statistics rather than the SIQR, since it is much easier to clothe the latter with meaning. From a set of data we identify the quartiles and then take the *midquartile,* $(Q_1 + Q_3)/2$, as a measure of location. (Note that the 'midquartile' is *not* to be understood as the second quartile or median, Q_2.) The SIQR is then a measure of dispersion, being the median of the absolute deviations from the midquartile.

There are four measures of location which each have a natural association with a particular measure of dispersion: the arithmetic mean with the standard deviation, the median with the mean deviation about the median, the mid-range with the range, and the midquartile with the semi-interquartile range. Each of these natural pairs has advantages and disadvantages which makes it appropriate to particular types of data.

Where the quartiles divide the observations into four equal parts, the *deciles* divide them into ten equal parts and the *centiles* (sometimes termed 'percentiles') divide them into 100 equal parts. We are often compelled to make use of the 1st and 9th deciles or of the 1st and 99th centiles because the extreme values are unascertainable.

We shall not encourage the reader to try to develop Program 5.1 so that it will print out deciles or centiles, but neither shall we seek to deter him.

5.2 Grouped data

Some writers and computer programmers have begun to use the term 'grouped data' to refer to frequency data such as Data 5 and Data 6 in Section 2.6. This is not what statisticians mean by the term.

Consider the following data:

Data 12

40, 43, 46, 59, 64, 67, 68, 69, 75, 76, 78, 80, 82, 82, 86, 90, 92, 127

These 18 observations take 17 different values; since 82 is the only value to occur more than once, it is the mode. The data would look rather silly if expressed as frequency data, but they could be conveniently *grouped* into six classes as follows

Class	*Frequency*
39–53	3
54–68	4
69–83	7
84–98	3
99–113	0
114–128	1

We might consider it more logical to 'group the data into groups' or to 'classify the data into classes', but statisticians have been 'grouping data into classes' from time immemorial.

When data are put into classes, we lose some of the information. If we then try to estimate the mean and variance from the grouped data, we will have lost some accuracy. But if information on thousands of observations has to be collected, recorded, and analysed, there are advantages in grouping; if a summary of the data has to be published, grouping may be inevitable. There was no need for grouping of the horse-kick data, Data 3, since although there were 280 observations there were only five different values among them. Grouping is necessary when both the number of observations and the number of different values are large.

We have not discussed how Data 12 arose. Were it not for the final observation the data could well have been examination marks. It could alternatively be the weights in kilograms of 18 adults, though the final observation is again rather exceptional. We would describe examination marks as *discrete data* and weights in kilograms as *continuous data.* Continuous data consist of observations on a continuous variable, that is, on a variable which can take any value — integral, fractional, or irrational — within its range. It can never be recorded 'exactly', and we usually infer that it has been rounded to the nearest multiple of the highest common factor of the recorded data. So if Data 12 were weights in kilograms we would infer that 40 means a weight of between 39.5 and 40.5 kg. This assumption determines the way in which we treat grouped data, and nothing is lost by applying the same practices even where the original data are known to be discrete.

Consider the following data:

Data 13

9.3	7.4	10.4	9.3	8.8	10.2	9.5	11.3	9.7	10.3
9.4	6.8	10.4	9.6	11.2	8.6	8.2	9.2	10.6	7.6
9.4	9.8	8.5	9.7	10.4	8.8	9.3	10.7	8.2	9.1
5.9	10.5	10.6	9.2	10.4	8.3	9.4	12.1	11.3	9.6

It is reasonable to infer that this is continuous data rounded to the nearest multiple of 0.1. It came to our notice in an examination question in which it purported to be the observed percentage moisture content of 40 samples of cotton. If we had tested 40 cartons each of which contained 10 boxes of matches, and had recorded for each carton the mean number of defective matches per box, we would similarly have obtained data consisting entirely of multiples of 0.1 but it would have been discrete data since the numbers of defective matches were integers and the mean of any 10 integers must be an exact multiple of 0.1.

In either case, we would choose the *class boundaries* for the purposes of grouping as odd multiples of 0.05, so that it would be impossible for an observation to lie on a class boundary. The smallest observation is 5.9 and the largest is 12.1. We could choose to have 8 classes with a *class interval* of 0.8 and hence the class boundaries would either be 5.85, 6.65, 7.45, . . ., 12.25, or 5.75, 6.55, 7.35, . . ., 12.15. For a class interval of 1.0 we would choose 5.55, 6.55, 7.55, . . ., 12.55, or 5.45, 6.45, 7.45, . . ., 12.45; it would be quite wrong to choose 4.95, 5.95, 6.95, . . ., 12.95, so that the integral part of the observation would determine the class, since the unecessary extra class (8 classes with class interval 1.0 when the range of the data is only 6.2) would give misleading results when the data were grouped.

Program 5.2 could be used to put the data into classes.

Program 5.2

```
 90 DIM N(255)
100 PRINT: PRINT: PRINT "    PUTTING DATA INTO CLASSES"
110 PRINT: INPUT "Input number of observations ";N
120 INPUT "Input lowest class boundary ";L
130 INPUT "Input highest class boundary ";H
140 INPUT "Input number of classes ";NC
150 PRINT: C = (H-L)/NC: PRINT "Class interval = ";C
160 FOR I = 1 TO N
170 PRINT "Input observation No. ";I;": ";
180 INPUT X: Y = INT((X-L)/C)
190 N(Y) = N(Y)+1
200 NEXT I
210 PRINT
220 PRINT "  =======                   ==========="
230 PRINT "  !RANGE!                   !FREQUENCY!"
240 PRINT "  =======                   ==========="
250 PRINT: K = L
260 FOR I = 0 TO NC-1
270 PRINT ;K;" TO ";K+C;TAB(30);N(I): K = K+C
280 NEXT I
290 END
```

Program notes

(1) If your BASIC will accept a declaration in terms of a variable, line 90 can be discarded and DIM N(NC−1) introduced after line 140.

(2) If at line 150 the calculated class interval is not an exact multiple of the highest common factor of the observations, an error has been made and the program should be re-commenced with a consistent set of proposals for L, H, and NC.

(3) At line 100 we have preferred the self-explanatory title to the more formal GROUPING OF DATA. At lines 220–240 we have

adopted another programmer's ornamentation, and might prefer the heading CLASS rather than RANGE.

The grouped data could be used as input for a program which is designed to accept data collected by frequencies. If so, the value taken as representative of each class would be the *class mid-interval,* calculated as: X = L + C*(I + 0.5) for the class whose frequency is N(I), for I = 0, 1, 2, . . ., NC − 1. Clearly the mean and the variance then calculated would be only an approximation to those of the original ungrouped data. There would be a tendency for the variance to be overstated; one reason for condemning the introduction of an unnecessary extra class is that it would aggravate this tendency.

If the class boundaries are 5.85, 6.65, 7.45, . . ., 12.25, it is perfectly acceptable to describe the classes by the *class limits* 5.9–6.6, 6.7–7.4, 7.5–8.2, . . ., 11.5–12.2. The class limits are the smallest and largest values which can fall into the class.

Before running Data 13 on Program 5.2 we modified line 270 by adding after N(I) the extra item '; TAB(28); L + C*(I + 0.5)' to print out the class mid-intervals, changing the TAB(30) to TAB(20). The output was then (putting L = 5.85, H = 12.25, NC = 8):

Range	*Frequency*	*Mid-interval*
5.85 to 6.65	1	6.25
6.65 to 7.45	2	7.05
7.45 to 8.25	3	7.85
8.25 to 9.05	5	8.65
9.05 to 9.85	15	9.45
9.85 to 10.65	9	10.25
10.65 to 11.45	4	11.05
11.45 to 12.25	1	11.85

Sometimes data have to be put into classes with unequal class intervals. For instance, if data on personal incomes were put into classes of equal size there would be either far too many classes for convenient presentation or far too large a proportion of the population in the class with the largest frequency for any useful information to be conveyed. Program 5.2 would need to be modified substantially in order to provide for unequal class intervals.

Continuous data can often be collected directly into class intervals. For instance, if the percentage moisture contents of samples of cotton were read from a scale — the moisture contents being determined from the conductivity or dielectric properties — and it were known that only grouped data were required, the operative could be instructed to produce a direct record of the

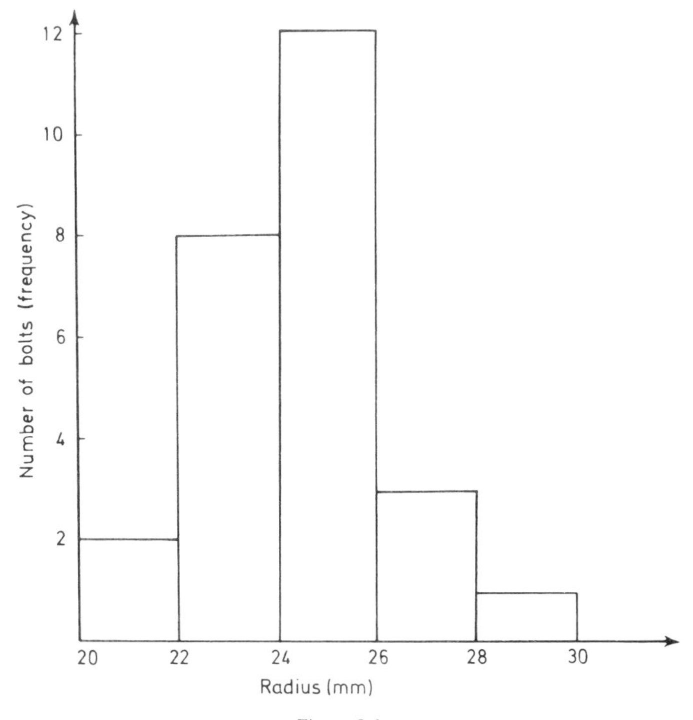

Figure 5.1

observed numbers of samples in each of the intervals 5.5–6.0, 6.0–6.5, etc. There would then be no need for a program to group the data. The same considerations apply to all data which are measured by analogue rather than digital instruments.

In Section 2.5 we indicated that a vertical bar chart is the best form of graphical representation of ungrouped discrete data. For grouped data, whether discrete or continuous, the best form of graphical representation is the *histogram*. Figure 5.1 is of the histogram of measurements of radii of 26 bolts; since the class boundaries are exact multiples of 2 mm it can be inferred that the measurements were collected directly into classes without being previously recorded more precisely.

We do not propose to enter into the complexities of printing out a histogram or a cumulative frequency diagram as an extension of Program 5.2. The task will be within the competence of the reader if he has mastered the 'graphics' of his computer, but it should be noted that the cumulative frequency diagram consists of straight lines of varying slopes and so requires high resolution graphics.

5.3 Moments of higher order

The *moment of order k about the origin* is denoted by m'_k and defined as:

$$m'_k = \frac{1}{N} \sum_{i=1}^{N} x_i^k$$

On comparing this with the formula for the arithmetic mean in Section 2.7, it will be seen that m'_1 is the arithmetic mean.

The *moment of order k about the mean* is denoted by m_k and defined as:

$$m_k = \frac{1}{N} \sum_{i=1}^{N} (x_i - \bar{x})^k$$

Again referring back to Section 2.7, it will be seen that m_2 is the mean squared deviation; that means it is $(N-1)/N$ times the variance.

Some writers, and the manufacturers of some pocket calculators, adopt the convention that the divisor $(N-1)$ is employed in calculating the 'sample standard deviation' but the divisor N is employed in calculating the 'population standard deviation'. It may be noted that the 'population standard deviation' under this absurd definition is the squareroot of m_2.

As was pointed out in Section 2.7, the mean of $(x_i - \bar{x})$ is always zero. Hence m_1 is always zero and never worth mentioning, and consequently there is unlikely to be confusion from the practice of using m as an alternative to m'_1 or $\bar{x}$ to represent the mean.

Referring yet again to Section 2.7, in which it was seen that the variance could be computed from the quantities which in Program 2.8 were summed as $S1$ and $S2$, it is only necessary to note that $S2/N$ is m'_2 in order to derive the identity

$$m_2 = m'_2 - m^2$$

The *third moment about the mean* (which is the way we would normally refer to 'the moment of order 3 about the mean') is an indication of *skewness*. By dividing m_3 by the cube of the squareroot of m_2 we obtain the *coefficient of skewness,* which is not only dimensionless but also unaffected by the relative variability of the data and so is sometimes described as a 'shape' statistic.

Just as the main interest in m'_2 was as a step in computing m_2 or the variance, so also the interest in m'_3 is mainly in order to use it for computing m_3 and the coefficient of skewness. The relevant identity proves to be:

$$m_3 = m'_3 - 3m'_2 m + 2m^3$$

For well-shaped data the median will always lie between the mean and the mode. These three measures of location are in alphabetical order from the left in the case of well-shaped data with negative skew (which is sometimes called 'skew to the left'), and in reverse alphabetical order for well-shaped data with positive skew ('skew to the right'). It has been demonstrated that for extremely well-shaped data the difference between the mean and the mode will be about three times the difference between the mean and the median. These considerations led to two versions of the Pearson Coefficient of Skewness:

$$\text{SK} = \frac{\text{Mean} - \text{Mode}}{\text{Standard Deviation}} \qquad \text{SK} = \frac{3(\text{Mean} - \text{Median})}{\text{Standard Deviation}}$$

Neither of these statistics is now used by serious statisticians. If we had a set of data with a pronounced positive skew but with the mode well above the mean (such as the salaries and wages in a company in which the only group of staff on *exactly* the same salaries are the working directors) the former version of SK would be extremely misleading; the latter version involves the inconvenience of having to compute the median.

The *fourth moment about the mean* is an indication of *kurtosis,* and may be computed from the moments about the origin using the identity:

$$m_4 = m'_4 - 4m'_3 m + 6m'_2 m^2 - 3m^4$$

All the relationships between m_k and the moments about the origin of order k or less can be obtained by expanding $(x_i - \bar{x})^k$ by the binomial theorem and separating out the $(k+1)$ terms to obtain $(k+1)$ summations involving decreasing powers of x_i in association with increasing powers of $\bar{x}$. The binomial coefficients when $k = 3$ are respectively 1, -3, 3, -1, and when $k = 4$ they are respectively 1, -4, 6, -4, 1, but in each case the last term cancels out part of the penultimate term to produce the identities we have presented.

The *coefficient of kurtosis* is m_4 divided by the square of m_2, and has the same properties of a 'shape' statistic as were specified in the case of the coefficient of skewness. It cannot be negative. For data whose histogram is a perfect rectangle its value is less than 2. For the 'bell-shape' of the normal distribution, which will be discussed in Section 7.3, its value is 3. A 'high peak and wide skirts' will produce a value of 5 or 6. Some statisticians take the normal distribution as a standard and so define the coefficient of kurtosis as $m_4/m_2^2 - 3$.

Distributions which then have a negative coefficient are given the eminently-forgettable label 'platykurtic' while those with a positive coefficient are labelled 'leptokurtic'.

A relative frequency curve which falls away from left to right, such as the 'horse-kick' Data 3 in Section 2.5, will always have a positive coefficient of skewness. If it falls away 'exponentially' it will also have a large coefficient of kurtosis; for the true 'negative exponential' curve which will be discussed in Section 7.2 the coefficient of kurtosis is 9, a value hardly ever attained by real data.

5.4 A descriptive statistics package

Program 5.3 is designed to provide a 'package' in descriptive statistics. Sections providing new options can be added as they are written, so that eventually a long program is produced offering a comprehensive service. The list of options is often called a 'menu'. So that the program can be used while it is not fully developed, and so that it can easily be understood, little attempt is made to economize on the length of the program by using common subroutines. In Chapter 6 we shall offer seven programs which can all run separately but can be put together as a single package. In Chapter 7 we shall deliberately go to the opposite extreme by presenting a single program which offers five options but is completely integrated so as to minimize the total length of the program.

A principal purpose of Program 5.3 is to offer the 'higher moments' and the shape statistics discussed in the previous section.

Program 5.3

```
 100  N = 0: S1 = 0: S2 = 0: S3 = 0: S4 = 0
 110  PRINT: PRINT: PRINT "                    STATISTICS"
 120  PRINT: PRINT "   1  Individual observations"
 130  PRINT: PRINT "   2  Data collected by frequencies"
 140  PRINT: PRINT "  Press 1 or 2 according to the form of
      the data"
 150  A = VAL(GET$)
 160  ON A GOTO 1000,2000
1000  PRINT: PRINT " Do you want the data to be grouped?
      (Y/N) "
1010  A$ = GET$
1020  IF A$="Y" GOTO 5000
1030  PRINT: PRINT " Do you know how many observations there
      are?  (Y/N) "
1040  A$ = GET$
1050  IF A$="Y" GOTO 1200
1060  PRINT: PRINT "Input the observations."
1070  PRINT: PRINT "Input 1E+18 at the end of data."
1080  INPUT X: IF X>1E+17 GOTO 3000
1090  N = N+1: S1 = S1+X: XX = X*X: S2 = S2+XX: S3 = S3+XX*X
      : S4 = S4+XX*XX
1100  GOTO 1080
```

```
1200  PRINT: INPUT "  Input the number of observations: ";N
1210  FOR I = 1 TO N
1220  PRINT: INPUT "   Input an observation: ";X
1230  S1 = S1+X: XX = X*X: S2 = S2+XX: S3 = S3+XX*X
      : S4 = S4+XX*XX
1240  NEXT I
1250  GOTO 3000
2000  PRINT: PRINT "   Are the intervals equal?  (Y/N) "
2010  T$ = GET$
2020  IF T$="Y" GOTO 2200
2030  PRINT: PRINT " Input each value, then its frequency."
2040  PRINT " Input 1E+18,1 when all the data is in."
2050  INPUT X,F: IF X>1E+17 GOTO 3000
2060  N = N+F: S1 = S1+F*X: XX = X*X: S2 = S2+F*XX
2070  S3 = S3+F*XX*X: S4 = S4+F*XX*XX
2080  GOTO 2050
2200  PRINT: INPUT "Input the minimum value and the
      interval: ";MN,C
2210  PRINT: PRINT " Input each frequency."
2220  PRINT " Input -1 when all the data is in."
2230  PRINT: X = MN
2240  PRINT X;" ";: INPUT F: IF F<0 GOTO 3000
2250  N = N+F: S1 = S1+F*X: XX = X*X: S2 = S2+F*XX
2260  S3 = S3+F*XX*X: S4 = S4+F*XX*XX
2270  X = X+C: GOTO 2240
3000  PRINT: PRINT: PRINT "Number of observations = ";N
3010  M = S1/N: PRINT: PRINT "Mean = ";M
3020  SS = S2-S1*M: V = SS/(N-1)
3030  PRINT "Variance = ";V
3040  S = SQR(V): PRINT "Standard deviation = ";S
3050  CV = S/M: PRINT: PRINT "Coefficient of Variation = ";
      CV;" = ";CV*100;"%"
3060  K2 = S2/N: K3 = S3/N: K4 = S4/N
3070  PRINT: PRINT "2nd moment about the origin = ";K2
3080  PRINT "3rd moment about the origin = ";K3
3090  PRINT "4th moment about the origin = ";K4
3100  PRINT: PRINT: PRINT "Press C to continue."
3110  A$ = GET$
3120  M2 = SS/N: PRINT: PRINT: PRINT "2nd moment about the
      mean = ";M2
3130  MM = M*M: M3 = K3-3*K2*M+2*MM*M: PRINT "3rd moment
      about the mean = ";M3
3140  M4 = K4-4*K3*M+6*K2*MM-3*MM*MM: PRINT "4th moment
      about the mean = ";M4
3150  CS = M3/M2/SQR(M2): PRINT: PRINT "Coefficient of
      Skewness = ";CS
3160  CK = M4/M2/M2: PRINT: PRINT "Coefficient of
      Kurtosis = ";CK
3170  PRINT: PRINT "Do you want to re-input the data to find"
3180  PRINT "the mean deviation, etc.?  (Y/N) "
3190  A$ = GET$
3200  IF A$="N" GOTO 100
3210  PRINT: PRINT: NN = 0: S1 = 0: S2 = 0: S3 = 0: S4 = 0
3220  IF A=2 GOTO 3400
3230  FOR I = 1 TO N
3240  PRINT: INPUT "  Input an observation: ";X
3250  D = X-M: S1 = S1+ABS(D): DD = D*D: S2 = S2+DD
3260  S3 = S3+DD*D: S4 = S4+DD*DD
3270  NEXT I
3280  GOTO 4000
3400  IF T$="Y" THEN X = MN-C
3410  IF T$="Y" THEN X = X+C: PRINT X;" ";: INPUT F
3420  IF T$="N" THEN PRINT "Input each value, then its
      frequency": INPUT X,F
3430  NN = NN+F: D = X-M: S1 = S1+F*ABS(D): DD = D*D
      : S2 = S2+F*DD
3440  S3 = S3+F*DD*D: S4 = S4+F*DD*DD
```

```
3450  IF NN<N GOTO 3410
4000  PRINT: PRINT "Mean Deviation = ";S1/N
4010  PRINT: PRINT "2nd moment about the mean = ";S2/N
4020  PRINT "                    Compare: ";M2
4030  PRINT: PRINT "3rd moment about the mean = ";S3/N
4040  PRINT "                    Compare: ";M3
4050  PRINT: PRINT "4th moment about the mean = ";S4/N
4060  PRINT "                    Compare: ";M4
4070  PRINT: PRINT "Coefficient of Skewness = ";
      S3/S2/SQR(S2/N)
4080  PRINT "                    Compare: ";CS
4090  PRINT: PRINT "Coefficient of Kurtosis = ";S4*N/S2/S2
4100  PRINT "                    Compare: ";CK
4110  PRINT: PRINT: PRINT " Do you want the median and mode?
      (Y/N) "
4120  A$ = GET$
4130  IF A$="Y" THEN PRINT: PRINT "  HARD LUCK!"
4140  GOTO 100
5000  PRINT: PRINT: PRINT "  HARD LUCK!": GOTO 1030
9999  END
```

Program notes

(1) Lines 110–160 offer the options and make provision for the appropriate section of the program to be called. There is plenty of room in the line-numbering and on the screen to offer further options later. All sections will commence with a line number which is a multiple of 1000.

(2) Lines 150–160 may give trouble. If your computer will not accept any version of GET or INKEY$, you may have to use INPUT A and double the number of key depressions required (alas!); if so, change 'Press' in line 140 to 'Input'. If your computer will not accept the ON A statement, try GOTO 1000*A; in the last resort you can always put IF A = 1 GOTO 1000, etc., although if your computer is *so* awkward it will probably insist on THEN before GOTO.

(3) In lines 1070–1080 we have assumed that the computer will accept the number 1E + 18. In some of our other programs we say Input 1E + 11 since it is quicker to input 11 than to input 18.

(4) In lines 1090, 1230, 2070, 2260, and 3060, the variables S1, S2, S3, S4 sum the respective powers of the observations. Putting XX = X*X improves the efficiency, and later we use MM and DD for the same purpose. M, K2, K3, K4 then become the first four moments about the origin, and are printed out in lines 3010 and 3070–3090.

(5) In lines 3120–3160, M2, M3, M4 are moments about the mean and are computed from the moments about the origin using the identities given in the text.

(6) In line 3220, A = 2 is recalling the type of data decided at lines 120-150.

(7) In lines 3250–3260, S1 is now re-defined as the sum of the

absolute deviations, and S2, S3, S4 become the sums of powers of the deviations.

(8) As the Program stands, we have to input the data twice in order to compute the mean deviation (though see the PROBLEMS regarding this). The program uses the deviations calculated at this time to produce revised calculations of the moments about the mean and compare them at lines 4010–4100 with the previous calculations. If we have made no mistake and there are no problems with the data, the results will show no significant discrepancies; the values previously calculated will usually be the most accurate ones if there are slight discrepancies. If all the values show marked discrepancies, we will almost certainly have put in an incorrect observation on either the first or the second batch of inputs; so 'having to input the data twice' is being used as a way of verifying the correctness of the input. But sometimes there will be serious discrepancies due to problems with the data of the types discussed in Section 3.3; the program may well have produced a DATA ERROR at line 1090, 1230, etc., and one must find ways to bypass the summing of the third and fourth powers of the observations and rely on lines 3430–3440 to calculate the higher moments.

(9) At line 4130 it is 'HARD LUCK' that the program will not as yet calculate the median and mode. But all the ideas have already been put forward in Program 5.1.

(10) At line 5000 it is 'HARD LUCK' that the program will not as yet put individual observations into classes. But all the ideas have already been put forward in Program 5.2.

PROBLEMS

(5.1) Attempt some or all of the following possible improvements to Program 5.1: (i) No mode is printed out if FH = 1; (ii) if X is less than MN or greater than MX, attempts are made to move the data up or down the array N with appropriate adjustments to MN or MX rather than immediately declaring DATA ERROR; (iii) to find the quartiles, the midquartile, and the SIQR; (iv) to find deciles and centiles; (v) to accept non-integer data; (vi) to accept data collected by frequencies; (vii) to the READ . . . DATA mode, employing also RESTORE, abandoning the array N; (viii) to print out a vertical bar chart; (ix) to print out a cumulative frequency diagram.

(5.2) Modify Program 5.2 so that the user can return to line 120 if not satisfied with the class interval printed out at line 150. Better still, develop it so that it will determine the class boundaries when given the minimum and maximum observations, the highest common factor of the observations, and the number of classes desired. Incorporate the suggested modification at line 270 to calculate the class mid-intervals, so that the output can be used as input to Program 5.3.

(5.3) Run the 'crushing loads data' from Problem 2.9 through Program 5.2, with class boundaries 4505.5, 4605.5, 4705.5, . . ., 5705.5. Run it through again with 10 classes instead of 12.

(5.4) Run the 'metal shaft data' from Problem 2.9 through Program 5.2 with class boundaries 2.9885, 2.9965, 3.0045, . . ., 3.0525. Run it through again with 7 classes instead of 8, trying to make L and H equidistant from the mid-range.

(5.5) Notes (8), (9) and (10) to Program 5.3 suggest possible improvements to handle different types of data and produce additional results, with Programs 5.1 and 5.2 amalgamated in. The enlarged program could also produce both forms of the Pearson Coefficient of Skewness. Attempt this enlarged program, preferably making it able to cope with awkward data such as Data 10 and Data 11.

(5.6) When Program 5.3 has been developed as much as intended, use it to find all the statistics for Data 2 to Data 7 from Chapter 2, Data 12 and Data 13 from Chapter 5, and the grouped data from Problems 5.3 and 5.4. Also find the statistics for the ungrouped data as in Problem 2.9 and compare them with those of the grouped data.

(5.7) The US Treasury Bulletin for April 1976 gives the 'amount (in millions of dollars) of marketable government bonds with various times to maturity in the years 1966, 1970, and 1975':

Year	*Up to 1 year*	*1–4 years*	*5–9 years*	*10–19 years*	*20–30 years*
1966	89 136	60 933	33 596	8 439	17 023
1970	105 530	89 615	15 882	10 524	11 048
1975	163 947	101 918	26 831	14 508	8 402

Observe that the class intervals are incoherent; bonds whose time to maturity is 0.5–1 year belong to both the first and the second class! Make illogical assumptions as to where the class mid-intervals are (logical assumptions being impossible) and find for each of the three dates the mean and variance of the time to maturity. Here is the second problem: which numbers are observations and which numbers are frequencies in using Program 5.3?

Chapter 6

Random variables and probability distributions

6.1 The concept of a random variable

At a leading university, students admitted to the undergraduate courses in mathematics or statistics were asked to tick such topics out of about fifty topics as they had already been taught at school. About half of them ticked 'binomial distribution', but scarcely 5 per cent ticked 'discrete random variables'. Yet it is impossible to be taught anything meaningful about the former topic without first being taught the latter. Subsequent questioning confirmed the state of muddle in the minds of students.

A *random variable* is a numerically-valued variable (perhaps better described as a *function*) defined on a sample space in such a way that each sample point corresponds to one and only one value of the variable.

In Chapter 4 we discussed the experiment of throwing a die, in which the sample space was $\{1,2,3,4,5,6\}$. We now only need to define the random variable X as 'the number thrown on the die', and we have produced a simple example; X can take the values 1, 2, 3, 4, 5 or 6.

When in Section 4.5 we discussed the 'sum of values' in the experiment of throwing two dice and listed the number of sample points for each of the integers 2 to 12, we were in effect discussing the set of possible values of a random variable.

It is sometimes helpful to think of a random variable as a 'partitioning' of the sample space according to a prescribed formula. In the experiment of throwing an unbiased die twice we defined the event B as 'sum of values thrown is 7' and the event C as 'sum of values thrown is 8'. If X is defined as 'the sum of the values thrown' then the event $X=7$ is identical to the event B and the event $X=8$ is identical to the event C. (Note accordingly that a random variable is not an event.) There are a further 9 events similar to B and C, characterized by the sum of the values thrown, which are all mutually exclusive; the 11 events together constitute the whole sample space, while separately they constitute a partitioning of the sample space according to the sum of the values thrown.

A *discrete sample space* consists either of a finite number of sample points or of a 'countably infinite' set of sample points such as the set of all positive integers. It can be contrasted with a continuous sample space which contains an infinite number of sample points in any finite interval of the space; observations on a continuous sample space give rise to continuous data as discussed in Section 5.2. Continuous sample spaces and continuous random variables will be discussed in Chapter 7.

Any random variable defined on a discrete sample space is a *discrete random variable.*

The probability of an event is the sum of the probabilities of the outcomes which it comprises. In the case of the unbiased die thrown twice discussed in Sections 4.4 and 4.5 we were told that the 36 sample points were all equally likely, and so the probabilities of the 11 events into which the sample space was partitioned by the random variable X = 'the sum of the values thrown' were 1/36 times the numbers of sample points which they comprised.

The *probability distribution* of a discrete random variable is the set of values which the random variable can take together with the associated probabilities. The probability distribution of X is therefore:

r:	2	3	4	5	6	7	8	9	10	11	12
$P(X=r)$:	$\frac{1}{36}$	$\frac{2}{36}$	$\frac{3}{36}$	$\frac{4}{36}$	$\frac{5}{36}$	$\frac{6}{36}$	$\frac{5}{36}$	$\frac{4}{36}$	$\frac{3}{36}$	$\frac{2}{36}$	$\frac{1}{36}$

On the same sample space we could define Y = 'difference between the values thrown' (i.e. larger value minus smaller value, irrespective of which was thrown first); there are six sample points for which $Y = 0$, ten sample points for which $Y = 1$, etc., the probability distribution being most conveniently expressed with 18 as the common denominator:

s:	0	1	2	3	4	5
$P(Y=s)$:	$\frac{3}{18}$	$\frac{5}{18}$	$\frac{4}{18}$	$\frac{3}{18}$	$\frac{2}{18}$	$\frac{1}{18}$

For the experiment of selecting two cards at random without replacement from a 52-card pack, the probability distribution of X= 'the number of aces obtained' is given by the fact that there are 48×47 sample points in the event $X=0$, (4×48) in the event 'only first card is an ace', (48×4) in the event 'only second card is an ace', and (4×3) in the event $X=2$. There are 52×51 sample points in the sample space, and so the probability distribution of X is:

r:	0	1	2
$P(X=r)$:	$\frac{188}{221}$	$\frac{32}{221}$	$\frac{1}{221}$

The probability distribution of Y = 'the number of spades obtained' is:

s:	0	1	2
$P(Y=s)$:	$\frac{19}{34}$	$\frac{13}{34}$	$\frac{2}{34}$

A probability distribution may be represented graphically by a *vertical bar chart* on the same lines as proposed for discrete data collected by frequencies in Program 2.6 (extension). It is left to the reader to make similar extensions to the programs for discrete probability distributions to be presented in this chapter.

The *cumulative distribution function* of the random variable X is denoted by $F(x)$ and defined as $P(X \leqslant x)$; that is, it is the probability that the random variable is less than or equal to any specified number x, expressed as a function of x. Note this distinction betwen the use of the capital letter for the random variable and the small letter for the values which it may take; we can perform algebra or calculus on x, but not on X. Note also that this definition of $F(x)$ applies to a continuous as well as to a discrete random variable.

For a discrete random variable taking only integral values, $F(r)$ may be obtained by summing the cumulative probabilities:

r:	0	1	2
$P(X=r)$:	$\frac{188}{221}$	$\frac{32}{221}$	$\frac{1}{221}$
$F(r)$:	$\frac{188}{221}$	$\frac{220}{221}$	$\frac{221}{221}$

and in general we may write

$$F(r) = \sum_{t=-\infty}^{r} P(X = t)$$

However, $F(x)$ is defined for all real values of x even though X takes only integral values. The full description of $F(x)$ for the above case is:

$$F(x) = \begin{cases} 0, & x < 0 \\ \frac{188}{221}, & 0 \leqslant x < 1 \\ \frac{220}{221}, & 1 \leqslant x < 2 \\ 1, & x \geqslant 2 \end{cases}$$

If X and Y are two discrete random variables defined on the same sample space, each taking only integral values, they are said to be *independent* if and only if:

$$P(X = r \cap Y = s) = P(X = r)\,P(Y = s) \quad \text{for } \textit{all } r, s$$

The fuller definition of *independent random variables,* applicable equally to discrete or continuous random variables, is that X and Y are independent if and only if:

$$P(X \leqslant x \;\cap\; Y \leqslant y) = P(X \leqslant x)\,P(Y \leqslant y) \quad \text{for all real } x, y$$

A clear knowledge and understanding of the different concepts 'independent events', 'independent trials', 'independent random variables' is *basic* to any sound understanding of elementary probability.

6.2 Mean and variance of a probability distribution

In Section 2.7 we defined the mean and variance of a set of N observations as (using m as an alternative for $\bar{x}$):

$$m = \frac{1}{N}\sum_{i=1}^{N} x_i \qquad\qquad s^2 = \frac{1}{N-1}\sum_{i=1}^{N}(x_i - \bar{x})^2$$

Program 2.8 (development) and some later programs accepted data collected by frequencies. If the different values are then denoted by x_j and their respective frequencies by f_j, and there are h different values in all, we may equate the sum of f_j over the h different values to the number of observations denoted by N. The formulae become:

$$m = \frac{1}{\sum_{j=1}^{h} f_j}\sum_{j=1}^{h} f_j x_j = \sum_{j=1}^{h}\frac{f_j}{\sum_{j=1}^{h} f_j}\, x_j$$

$$s^2 = \frac{1}{\left(\sum_{j=1}^{h} f_j\right)-1} \sum_{j=1}^{h} f_j (x_j - \bar{x})^2 = \sum_{j=1}^{h} \frac{f_j}{\left(\sum_{j=1}^{h} f_j\right)-1} (x_j - \bar{x})^2$$

As the number of observations increases, all possible values appear among the data and the relative frequencies tend towards the limiting relative frequencies, which are the probabilities of the form $P(X = x_j)$ where X is the discrete random variable on which observations are being made. The effect of the (-1) in the formula for s^2 disappears as the limit is approached. The limiting form of m and s^2 are the mean and variance of the probability distribution of X, and are denoted by the equivalent Greek letters μ and σ^2. So the formulae become:

$$\mu = \sum P(X = x_j) x_j \qquad \sigma^2 = \sum P(X = x_j)\,(x_j - \mu)^2$$

where in each case the summation is over all possible values of x_j.

For a random variable X taking only integral values we may write:

$$\mu = \sum_{r=-\infty}^{\infty} rP(X = r) \qquad \sigma^2 = \sum_{r=-\infty}^{\infty} (r-\mu)^2 \; P(X = r)$$

Expanding $(r - \mu)^2$ in the last term, we obtain an identity corresponding to that obtained in Section 2.7 for s^2, and equally useful.

$$\sigma^2 = \sum_{r=-\infty}^{\infty} r^2 P(X = r) \; - \; \mu^2$$

In the standard probability distributions to be discussed shortly, there will be simple formulae for μ and σ^2 based on the numerical characteristics of the distributions which are termed *parameters*. It is however useful to show for simple distributions how μ and σ^2 can readily be computed from the above formulae. It is simplest to multiply all the probabilities by their lowest common denominator.

For the number thrown on an unbiased die:

r:	1	2	3	4	5	6	*Sum*
$6P(X = r)$:	1	1	1	1	1	1	6
$6rP(X = r)$:	1	2	3	4	5	6	21
$6r^2P(X = r)$:	1	4	9	16	25	36	91

$$\mu = \frac{21}{6} = 3\frac{1}{2} \qquad \sigma^2 = \frac{91}{6} - \left(\frac{7}{2}\right)^2 = \frac{182-147}{12} = \frac{35}{12}$$

For the number of aces obtained in selecting two cards at random without replacement:

r:	0	1	2	*Sum*
$221P(X=r)$:	188	32	1	221
$221rP(X=r)$:	0	32	2	34
$221r^2P(X=r)$:	0	32	4	36

$$\mu = \frac{34}{221} = \frac{2}{13} \qquad \sigma^2 = \frac{36}{221} - \left(\frac{2}{13}\right)^2 = \frac{468-68}{2873} = \frac{400}{2873}$$

6.3 The Bernoulli distribution

It is intended to present programs for calculating seven different types of probability distribution. When we speak of 'the binomial distribution' we are really referring to a particular *type* of probability distribution, applicable to a discrete random variable defined in a specified manner on a specified type of sample space. The actual distribution arising from a particular experiment will depend on one or more numerical facts appertaining to that experiment which determine the parameters of the distribution. Textbooks often define the word 'parameter' as a 'numerical fact about a population', by analogy with the definition of a 'statistic' as a 'numerical fact about a sample'; but it is more commonly applied to probability distributions (which may be regarded as infinite populations of potential observations).

The seven distributions will be presented as separate programs, but with line numbers such that all the programs can readily be combined into a single 'discrete probability distributions' package. The first program incorporates the 'menu' for the package. This lists the seven distributions in alphabetical order, but fortunately this is a perfectly convenient order in which to discuss them.

The simplest distribution is the *Bernoulli distribution,* which is the distribution of a random variable defined on a sample space which has only two outcomes. It does not matter what those outcomes are, but typical sample spaces which will readily suggest the kind of experiment to which they appertain are

{heads, tails}   {sound, defective}   {boy, girl}

{hit, miss}   {accepted, rejected}   {fatality, recovery}

In some of these cases there is an obvious hint of a dichotomy between a 'failure' and a 'success'. We define X as the number of successes, and so X must take the value 0 or 1. However, it is customary to use the word 'success' for whichever of the outcomes we wish to count. A quality controller is usually trying to locate any defective item, and so with surprising frequency the outcome 'sound' is equated to 'failure' and 'defective' is equated to 'success'. In discussing the experiment of having a baby whose sex is not predetermined, the kindly male lecturer to a class of students of whom most but not all are male will always define boy as 'failure' and girl as 'success'.

The symbol p is used to denote the probability of 'success', and this one parameter defines the distribution completely. Obtaining the mean and variance is simple, following the procedure adopted in the previous section:

r:	0	1	*Sum*
$P(X=r)$:	$1-p$	p	1
$rP(X=r)$:	0	p	p
$r^2P(X=r)$:	0	p	p

$$\mu = p \qquad \sigma^2 = p-p^2 = p(1-p)$$

Program 6.1 contains the 'menu' together with the program for the Bernoulli distribution. It may be noted that the 'moments about the origin' and the 'moments about the mean' are defined for probability distributions in an analogous manner to those for observed data, and denoted by μ'_k and μ_k respectively. It is easy to see that for the Bernoulli distribution the moments about the origin will be equal to p for all values of k. However, we have not attempted to produce moments of higher order for the probability distributions in this text, since their main function is to help in identifying the type of probability distribution which 'fits' a set of data and this is outside the scope of the text.

It is customary to use q to denote the probability of a 'failure' in this type of simple experiment, which is termed a 'Bernoulli trial'. Then $q=1-p$ and $\sigma^2=pq$.

Program 6.1

```
100 REM DISCRETE PROBABILITY DISTRIBUTIONS
110 PRINT: PRINT: PRINT " 1 BERNOULLI DISTRIBUTION"
120 PRINT: PRINT " 2 BINOMIAL DISTRIBUTION"
130 PRINT: PRINT " 3 GEOMETRIC DISTRIBUTION"
140 PRINT: PRINT " 4 HYPERGEOMETRIC DISTRIBUTION"
150 PRINT: PRINT " 5 NEGATIVE BINOMIAL DISTRIBUTION"
160 PRINT: PRINT " 6 POISSON DISTRIBUTION"
170 PRINT: PRINT " 7 UNIFORM DISCRETE DISTRIBUTION"
200 PRINT: PRINT: PRINT " Which kind of distribution do"
210 PRINT " you want to study?"
220 PRINT: PRINT "  Press 1,2,3,4,5,6, or 7."
230 D = VAL(GET$)
240 ON D GOTO 1000,2000,3000,4000,5000,6000,7000
```

```
1000 REM BERNOULLI DISTRIBUTION
1100 PRINT: PRINT: PRINT "    BERNOULLI DISTRIBUTION"
1110 PRINT: PRINT "X is the number of successes in one"
1120 PRINT "Bernoulli trial which has probability"
1130 PRINT "p of success."
1140 PRINT: INPUT "Do you want to input p as a fraction?
     (Y/N) ";A$
1150 IF A$="N" THEN PRINT: INPUT "Input p ";P: GOTO 1170
1160 PRINT: INPUT "Input numerator then denominator ";A,B
     : P = A/B
1170 G = 1-P: V = P*G
1200 PRINT: PRINT "Mean = p = ";P
1210 PRINT "Variance = ";V
1220 PRINT "Standard Deviation = ";SQR(V)
1300 W = G: Y = W
1400 PRINT
1420 PRINT: PRINT " r    P(X=r)";TAB(21);"P(X<=r)"
1430 PRINT
1440 FOR I = 0 TO 1
1460 PRINT " ";I;"   ";W;TAB(20);Y
1480 W = P: Y = Y+W
1490 NEXT I
1510 PRINT: PRINT "Press R to return."
1520 R$ = GET$
1530 IF R$="R" GOTO 1100
```

Program notes

(1) Lines 110–210 offer the ‘menu’, with room to add one or two distributions, but there is no point in introducing lines 100–240 to the computer until you want to put in more than one distribution.

(2) If you have trouble with lines 230–240, see note (2) to Program 5.3.

(3) It will be our policy to ensure that there is a printout similar to lines 1100–1130 for each distribution, indicating the type of experiment and random variable for which the distribution is appropriate. Anyone who does not understand the ‘explanation’ can still operate the program but is clearly incompetent to interpret the results.

(4) The option to input p as a fraction could be inserted into one or two of the other programs if desired. It can be useful if the sample space has been produced by condensing a more fundamental sample space, such as treating the outcome ‘six’ on one throw of a die as a ‘success’ which has probability 1/6 and treating any other value as a ‘failure’.

(5) Lines 1200–1220 have been designed to standardize with lines 2200–2220, 3200–3220, etc., in subsequent programs, and the attempt at similar standardization makes some of the lines 1300–1490 appear a little odd. The aim is to make it easier to compare the logic of the different programs, to copy lines from one program to another, or even to amalgamate identical sections into a shared subroutine.

6.4 The binomial distribution

If one is concerned to count the number of successes in a sequence of n trials, each of which has probability p of success and is independent of all the other trials, then the random variable which is of interest has a *binomial distribution*. The parameters are n and p, and it is a common shorthand to write '$X \sim b(n,p)$' to mean 'the random variable X has a binomial distribution with parameters n and p'. If we write simply '$X \sim b(2,\tfrac{1}{2})$' there can be no confusion as to which parameter is which; this would be the correct information for X = 'the number of heads thrown' in the experiment of tossing a coin twice, provided that the coin was unbiased and the trials were independent.

For the general case where $X \sim b(2,p)$ we have:

r:	0	1	2	*Sum*
$P(X=r)$:	q^2	$2qp$	p^2	1
$rP(X=r)$:	0	$2qp$	$2p^2$	$2p$
$r^2P(X=r)$:	0	$2qp$	$4p^2$	$2p(1+p)$

Here we have retained the convention as for the Bernoulli distribution that $q = 1-p$. The probability distribution arises very simply by applying the Law of Multiplication for independent events taking account of the fact that the trials are independent. Clearly

$$2qp+2p^2 = 2p(q+p) = 2p, \text{ and}$$
$$2qp+4p^2 = 2p(q+2p) = 2p(1+p).$$

Then $\mu = 2p$ and $\sigma^2 = 2p(1+p)-(2p)^2 = 2p(1-p) = 2pq$.

In general, the binomial distribution is the distribution of the sum of n independent random variables each of which has a Bernoulli distribution with parameter p. It follows that a Bernoulli distribution is simply a $b(1,p)$ distribution. One may also say that the binomial distribution is the distribution of the number of successes in n identically-distributed Bernoulli trials; a Bernoulli trial is simply a trial which has only two possible outcomes, but the definition of a sequence of Bernoulli trials includes the requirement that the trials are all independent of each other.

It is left to the reader to confirm, with the help of his statistics textbook if necessary, that the general formula for the probability distribution is given by the term involving $q^{n-r}p^r$ in the binomial series expansion of $(q+p)^n$. This can either be argued from the Law of Multiplication for independent events together with the formula for ${}_nC_r$, or derived from the Bernoulli distribution using probability generating functions — which will not be discussed in this text. (Note however that merely expanding the binomial expression does nothing to prove its relevance to the random variable under consideration — a fact which thousands of examination candidates every year seem to overlook.)

Most textbooks also offer a derivation of the mean and variance of the distribution, either avoiding or employing probability generating functions. The general formulae are:

$$\mu = np \qquad \sigma^2 = npq$$

and it will be seen by putting $n = 1$ and then $n = 2$ that these formulae are consistent with the results already obtained for those two cases.

Relatively few textbooks discuss the ratio between $P(X=r)$ and $P(X=r+1)$ which is important for rapid computation of the whole distribution. Using the general formula:

$$P(X=r) = \frac{n!}{(n-r)!\,r!}\,q^{n-r}p^r,\ r=0, 1, \ldots, n$$

we can produce the corresponding formula for $P(X=r+1)$ and hence establish the relationship:

$$P(X=r+1) = \frac{n-r}{r+1}\,\frac{p}{q}\,P(X=r),\ r = 0, 1, \ldots, n-1$$

Actually the relationship remains true for $r \geqslant n$, but merely tells us that $P(X=r)$ is zero for $r > n$.

In Program 6.2 the relationship here derived is used to compute each term in the probability distribution from the previous term, having of course first put $P(X=0) = q^n$ which is an obvious result from the Law of Multiplication. We shall continue the practice of computing $P(X=r+1)$ from $P(X=r)$ in the ensuing programs for other distributions.

Another feature of Program 6.2 is the neat printout obtained by the subroutine at lines 2700–2770, using the constants defined at line 2010. To save space this subroutine has not been put into the other programs, but it could easily be copied into them if desired, together with line 2010. On the other hand, if the programs were all put together so that they could be called from the 'menu' then the subroutine at 2700–2770 could be called from all the other sections of the amalgamated program simply by copying lines 2460–2470 as lines 3460–3470, 4460–4470, etc., and transferring line 2010 to become line 90.

Program 6.2

```
2000 REM BINOMIAL DISTRIBUTION
2010 Z$ = "000000": CF = 0.0000005
2100 PRINT: PRINT: PRINT "    BINOMIAL DISTRIBUTION"
2110 PRINT: PRINT "X is the number of successes in n"
```

```
2120 PRINT "Bernoulli trials each of which has"
2130 PRINT "probability p of success."
2135 REM PUTTING n = 1 PRODUCES THE BERNOULLI DISTRIBUTION.
2136 REM IF n IS LARGE AND p IS SMALL, X HAS APPROXIMATELY
2137 REM A POISSON DISTRIBUTION WITH MEAN EQUAL TO n*p.
2140 PRINT: INPUT "Input n,p ";N,P
2150 Q = 1-P: FL = 0
2160 M = N*P: V = M*Q
2200 PRINT: PRINT "Mean = ";M
2210 PRINT "Variance = ";V
2220 PRINT "Standard Deviation = ";SQR(V)
2230 IF N*LOG(Q)<-18 GOTO 2600
2300 W = Q^N: PQ = P/Q
2310 Y = W
2320 RL = 0: RU = 4: IF N<RU THEN RU = N
2400 PRINT
2410 IF FL>0 THEN PRINT: PRINT "X is the number of FAILURES"
2420 PRINT: PRINT " r    P(X=r)";TAB(21);"P(X<=r)"
2430 PRINT
2440 FOR I = RL TO RU
2450 IF I<10 THEN PRINT " ";
2460 PRINT ;I;: A = W: GOSUB 2700: PRINT "   ";S$;
2470 A = Y: GOSUB 2700: PRINT TAB(20);S$
2480 W = W*(N-I)*PQ/(I+1): Y = Y+W
2490 NEXT I
2500 IF RU<N THEN PRINT: PRINT "Press C to continue."
2510 PRINT: PRINT "Press R to return."
2520 R$ = GET$
2530 IF R$="R" OR RU=N GOTO 2100
2540 RL = RU+1: RU = RU+12: IF N<RU THEN RU = N
2550 PRINT: PRINT: PRINT "    BINOMIAL DISTRIBUTION"
2560 PRINT: PRINT "     n = ";N;"  p = ";P
2570 GOTO 2400
2600 IF N*LOG(P)>-18 THEN W = P^N: PQ = Q/P: FL = 1
     : GOTO 2310
2610 PRINT: PRINT "This program will not handle such
     a large n when p = ";P
2620 GOTO 2100
2700 A = A+CF
2710 IF A>0.1 THEN S$ = LEFT$(STR$(A),8): RETURN
2720 FOR J = 1 TO 5
2730 A = 10*A: IF A>0.1 GOTO 2760
2740 NEXT J
2750 S$ = "0.000000": RETURN
2760 S$ = "0."+LEFT$(Z$,J)+MID$(STR$(A),3,6-J)
2770 RETURN
```

Program notes

(1) For the use of lines 2010, 2460–2470 and 2700–2770, and their applicability to other programs, see the comments before the program.

(2) The relationships between distributions, such as those specified at lines 2135–2137, can be checked using appropriate values for the parameters both to test the soundness of the programs and to assess the closeness of the approximations.

(3) Lines 2230 and 2600–2610 take account of the fact that the computer cannot handle numbers much smaller than 1E−18. Note that LOG means the logarithm to base 10; adapt these lines to suit your computer.

(4) Line 2600 arranges to interchange failures with successes if the

computer can then cope. The 'flag' variable FL is set to 1 so that line 2410 will provide the explanatory printout.

(5) Lines 2320, 2440–2450, and 2500–2570 arrange for a readable printout of 12 lines at a time. Similar lines will appear at corresponding positions in the other 'discrete probability distribution' programs.

(6) At line 2300, or at line 2600 if appropriate, W is set at $P(X=0)$. At line 2480, W is set at $P(X=I+1)$ using the relationship with $P(X=I)$ discussed earlier. Note that PQ had been set at P/Q at line 2300, or at Q/P at line 2600 if appropriate.

6.5 The geometric distribution

If one is concerned with a sequence of Bernoulli trials, as defined in the previous section, but the sequence is to terminate as soon as the first 'success' is obtained, then the random variable of interest is the number of trials up to and including the first success. This has a *geometric distribution,* whose only parameter is p which is again defined as the probability of success on any one trial.

Examples of the distribution often appear in the form 'toss a coin until the first head is obtained' or 'throw a die until the first six is obtained'. More practical examples are 'attempt to thread a needle repeatedly until the first success' or 'fire shells at an enemy ammunition dump until the first success' in which it is more reasonable to assume that the experiment will terminate at the first success; but it is somewhat unlikely that in these cases the required conditions for a geometric distribution that the trials be independent with constant probability of success (i.e., that there is no 'learning' involved, no improvement of the aim when it is seen where the earlier shells have fallen) will be fulfilled.

Since X is the number of trials up to and including the first success, it is not difficult to recognize that $P(X=1)$ must be the probability of a success on the first trial, which is p. The probability of a failure on the first trial and a success on the second trial is qp, using the Law of Multiplication for independent events, and so this is the probability that $X=2$. It is soon clear that the general formula is

$$P(X=r) = q^{r-1}p, \text{ for } r=1, 2, 3, \ldots,$$

the number of sample points being countably infinite. Each sample point gives rise to a different value of the random variable, so the sample points are not equally likely.

The successive terms in the probability distribution form a geometric sequence with common ratio q. The cumulative distribution function is seen to be $F(r) = 1 - q^r$, either by summing the first r terms of the probability distribution or from the Law of Complementarity using the fact that $X > r$ if and only if the first r trials all result in failures.

The mean is easily determined if one remembers the expansion:

$$(1-q)^{-2} = 1 + 2q + 3q^2 + \dots + r\,q^{r-1} + \dots$$

so that:

$$\mu = p + 2qp + 3q^2p + \dots + r\,q^{r-1}p + \dots$$

$$= p(1-q)^{-2} = \frac{1}{p}$$

More complicated manipulations of series, or the use of probability generating functions, lead to the formula $(1+q)/p^2$ for the second moment about the origin and hence:

$$\sigma^2 = \frac{1+q}{p^2} - \left(\frac{1}{p}\right)^2 = \frac{q}{p^2}$$

Sometimes the geometric distribution is defined as the number of Bernoulli trials *before* the first success. This simply moves the whole distribution down one place, so that $P(X=0)=p$, $P(X=1)=qp$, . . ., $P(X=r)=q^rp$, etc. All the probabilities can be determined from the original program and ascribed to the stepped-down values of r. The value of μ is reduced by 1 and so becomes q/p; the variance is unchanged.

Program 6.3

```
3000 REM GEOMETRIC DISTRIBUTION
3100 PRINT: PRINT: PRINT "     GEOMETRIC DISTRIBUTION"
3110 PRINT: PRINT "X is the number of Bernoulli trials up"
3120 PRINT "to and including the first success, each";
3130 PRINT "trial having probability p of success."
3135 REM THE DISTRIBUTION OF (X-1), THE NUMBER OF BERNOULLI
3136 REM TRIALS BEFORE THE FIRST SUCCESS, IS ALSO CALLED
3137 REM A GEOMETRIC DISTRIBUTION.
3140 PRINT: INPUT "Input p ";P
3150 Q = 1-P: M = 1/P: V = Q*M*M
3200 PRINT: PRINT "Mean = ";M
3210 PRINT "Variance = ";V
3220 PRINT "Standard Deviation = ";SQR(V)
3300 W = P
3310 Y = W: RL = 1: RU = 5
3400 PRINT
3420 PRINT: PRINT " r     P(X=r)";TAB(21);"P(X<=r)"
3430 PRINT
3440 FOR I = RL TO RU
3450 IF I<10 THEN PRINT " ";
3460 PRINT ;I;"    ";W;TAB(20);Y
3480 W = W*Q: Y = Y+W
```

```
3490 NEXT I
3500 PRINT: PRINT "Press C to continue."
3510 PRINT: PRINT "Press R to return."
3520 R$ = GET$
3530 IF R$="R" GOTO 3100
3540 RL = RU+1: RU = RU+12
3550 PRINT: PRINT: PRINT "     GEOMETRIC DISTRIBUTION"
3560 PRINT: PRINT "      p = ";P
3570 GOTO 3400
```

Program notes

Some of the Program notes for Programs 6.1 and 6.2 are relevant also to Program 6.3 and subsequent programs. There are no special notes for Program 6.3, which is a much simpler program than Program 6.2.

6.6 The hypergeometric distribution

Although the prefix 'hyper-' means 'extreme' or 'excessive', we can see no obvious connection between the hypergeometric distribution and the geometric distribution.

When a quality controller selects a sample of n specimens from a batch of N specimens of which S are defective, he will usually carry out his statistical analysis on the assumption that he is investigating a binomial distribution with parameters n and p. He will be trying to draw inferences about p, which is unknown. If the sample is of size 1 his assumption will be correct; he will be investigating a Bernoulli distribution with parameter S/N and presumably S will be unknown. If the sample is selected at random then for $n > 1$ the appropriate distribution will be the binomial distribution if he is sampling with replacement but if (as is more likely) he is sampling without replacement it will be the hypergeometric distribution. If N and S are both very large compared with n, there will be little difference between the two distributions — assuming of course that p is equated to S/N.

By analogy with the binomial distribution it is convenient (though not a common practice) to describe S as the 'number of successes in the population', where the batch constitutes the population. (The discerning reader will note from what we said about the use of the word 'parameter' that the word 'population' could also be used to refer to the infinite population of repetitions of the whole experiment. It is important to distinguish the two meanings.) If X is the number of 'successes' in the sample, then it is a relatively simple application of the principles expounded in Section 4.4 to see that:

$$P(X=r) = \frac{{}_SC_r \; {}_{N-S}C_{n-r}}{{}_NC_n}, \quad r = 0, 1, \ldots, n$$

Just as there are two different meanings for 'population', so there are three different conventional uses for the symbol n in probability theory. In Section 4.3 we adopted the usual convention that n is the number of points in the sample space. In the formulae for ${}_nP_r$ *and* ${}_nC_r$ referred to in Section 4.4, it is accepted that n is the size of the population (in the sense of 'finite batch') while r is the size of the sample and there are ${}_nP_r$ sample points. When we started to discuss probability distributions we adopted the third convention, that n is the size of the sample; in our present example, the size of the population is N and there are ${}_NP_n$ sample points in the sample space. We re-emphasize that all these uses are standard usages in probability and statistics, which is why statisticians repeatedly give warnings against the mindless learning of formulae. (Note also that there is no logical relationship between the word 'sample' and the term 'sample space', which is why one or two writers have striven — unsuccessfully — to introduce the term 'possibility set'.)

In deriving the formula for the hypergeometric probability distribution, we follow our own advice that 'problems in probability which involve combinations rather than permutations are often simplest dealt with by considering the numbers of equally-likely combinations without reference to the fundamental sample space in which each permutation is a sample point'.

For the experiment of selecting two cards at random without replacement from a 52-card pack, we have a hypergeometric distribution with $N=52$, $n=2$, and so the denominator in the formula is 1326. If 'success' is defined as 'ace', then $S=4$; if 'success' is defined as 'spade', then $S=13$. In either case, we obtain the appropriate probability distribution from the formula in agreement with the results presented in Section 6.1.

The formula may be expanded in terms of factorials:

$$P(X=r) = \frac{S!\,(N-S)!\,(N-n)!\,n!}{(S-r)!\,r!\,(N-S-n+r)!\,(n-r)!\,N!}$$

and by producing the corresponding formula for $P(X=r+1)$ we establish the relationship:

$$P(X=r+1) = \frac{(S-r)\,(n-r)}{(r+1)\,(N-S-n+r+1)}\,P(X=r)$$

The formula for $P(X=r)$ is restricted to the range $0 \leqslant r \leqslant n$. But in order that the probability should not be zero it is also necessary that $r \leqslant S$ and that $(n-r) \leqslant (N-S)$. The latter inequality means that if $n > N-S$ it is impossible to have zero successes; this is provided for in the formula for $P(X=r)$ since $(N-S)!/(N-S-n)!$ is interpreted as $(N-S)\,(N-S-1)\,.\,.\,.\,(N-S-n+1)$ which will include a zero as one of the terms to be multiplied together, but it means that we cannot start computing the probabilities by finding $P(X=0)$ and repeatedly using the relationship between $P(X=r+1)$ and $P(X=r)$.

This is the same problem as arose with the binomial distribution, except that $P(X=0)$ is actually zero rather than simply too small to be computed. Once again we adopt the policy of trying to count failures if we cannot count successes, but once again our program throws in the sponge if it cannot start work at $P(X=n)$ — in the case of the hypergeometric distribution, because $n > S$. This does not mean that it is impossible to compute the binomial distribution or the hypergeometric distribution in the cases where $P(X=0)$ and $P(X=n)$ are both either very small or zero; but to do so would require a more complicated program.

It is rather tedious to calculate the mean and variance of the hypergeometric distribution. The formula for μ is easy to accept, since it is simply n multiplied by the proportion of successes in the population. Writing this proportion as p, we see that the mean is the same as for a binomial distribution. Writing q as the proportion of failures in the population, so that $q=(N-S)/N=1-p$, it can be seen that for $n>1$ the variance is less than for the corresponding binomial distribution (that is, the distribution which would be applicable if we sampled with replacement instead of without replacement).

$$\mu = \frac{nS}{N}(=np) \qquad \sigma^2 = \frac{nS\,(N-S)\,(N-n)}{N^2\,(N-1)} \left(= npq\frac{N-n}{N-1}\right)$$

The results for the two-cards examples in Sections 6.1 and 6.2 may be checked against these formulae. It is also useful to put $n=1$ in all the formulae in this section and confirm that they fit exactly with the corresponding results for the Bernoulli distribution with $p = S/N$.

The important practical uses of statistics involve drawing random samples from populations far more often than tossing coins or throwing dice, and sampling is normally without replacement from finite batches. So the hypergeometric distribution is widely applicable, and with the availability of computers there is no excuse for not making proper use of it. (Note however that sampling without replacement from a production line may be regarded as sampling from an infinite population, and so the binomial distribution again becomes appropriate.)

Program 6.4

```
4000 REM HYPERGEOMETRIC DISTRIBUTION
4100 PRINT: PRINT: PRINT "   HYPERGEOMETRIC DISTRIBUTION"
4110 PRINT: PRINT "X is the number of successes in a"
```

```
4120 PRINT "sample of size n selected at random"
4130 PRINT "without replacement from a population"
4140 PRINT "of size N of which S are successes."
4145 REM IF N AND S ARE VERY LARGE AND n IS NOT LARGE, X HAS
4146 REM APPROXIMATELY A BINOMIAL DISTRIBUTION WITH p = S/N.
4150 PRINT: INPUT "Input n,N,S ";K,N,S
4160 F = N-S: FL = 0
4170 M = K*S/N: V = M*(N-K)*F/N/(N-1)
4200 PRINT: PRINT "Mean = ";M
4210 PRINT "Variance = ";V
4220 PRINT "Standard Deviation = ";SQR(V)
4230 IF K>F GOTO 4600
4300 W = 1: FOR I = 0 TO K-1: W = W*(F-I)/(N-I): NEXT I
4310 Y = W: U = F-K+1: T = K: IF S<T THEN T = S
4320 RL = 0: RU = 4: IF T<RU THEN RU = T
4400 PRINT
4410 IF FL>0 THEN PRINT: PRINT "X is the number of FAILURES"
4420 PRINT: PRINT " r    P(X=r)";TAB(21);"P(X<=r)"
4430 PRINT
4440 FOR I = RL TO RU
4450 IF I<10 THEN PRINT " ";
4460 PRINT ;I;"    ";W;TAB(20);Y
4480 W = W*(S-I)*(K-I)/(I+1)/(U+I): Y = Y+W
4490 NEXT I
4500 IF RU<T THEN PRINT: PRINT "Press C to continue."
4510 PRINT: PRINT "Press R to return."
4520 R$ = GET$
4530 IF R$="R" OR RU=T GOTO 4100
4540 RL = RU+1: RU = RU+12: IF T<RU THEN RU = T
4550 PRINT: PRINT: PRINT "   HYPERGEOMETRIC DISTRIBUTION"
4560 IF FL>0 THEN S = F
4570 PRINT: PRINT "      n = ";K;"  N = ";N;"  S = ";S
4580 GOTO 4400
4600 IF K<=S THEN S = F: F = N-S: FL = 1: GOTO 4300
4610 PRINT: PRINT "The case where n>N-S and n>S, so that"
4620 PRINT "it is not possible to obtain either"
4630 PRINT "zero successes or n successes, is not"
4640 PRINT "yet programmed.": GOTO 4100
```

Program notes

(1) All the general comments on the style and layout of this program have already been made in the notes to Programs 6.1 and 6.2, while the formulae applied have been fully discussed in the text. It remains only to reconcile the notation. At line 4170, M represents μ, K represents n, and F represents $(N-S)$. Provided that $n \leqslant N-S$ the value of $P(X=0)$ is assigned to W by line 4300. The relationship applied at line 4480 is that presented in the text, noting that U represents $(N-S-n+1)$.

(2) T is the maximum possible value of r which has a non-zero probability, and is therefore whichever is smaller of n or S. This is determined at line 4310. Lines 4320, 4500 and 4540 all ensure that the printing-out of the distribution terminates at T.

(3) If $n > N-S$ but $n \leqslant S$, the values of S and F are interchanged at line 4600. Putting the flag FL to 1 ensures that the correct printout is given at lines 4560–4570, but the statement *IF FL>O* THEN $S = N-F$ should be added afterwards.

6.7 The negative binomial distribution

The negative binomial distribution stands in the same relationship to the geometric distribution as does the binomial distribution to the Bernoulli distribution: it is the distribution of the sum of k independent random variables each of which has a geometric distribution with parameter p. It may also be defined as the distribution of the number of Bernoulli trials up to and including the kth success, where each trial has probability p of success (noting that that the concept of Bernoulli trials includes the requirement that the trials are all independent of each other). It is convenient to use p yet again as a parameter, since it has always been given the same meaning, and we again use $q=1-p$. The total number of trials has now become the random variable whereas for the binomial distribution it was a parameter; in exchange, the number of successes has ceased to be a random variable and has become the parameter k.

The geometric distribution may now be seen to be a negative binomial distribution in which the parameter k takes the value 1.

The name 'negative binomial distribution' arises from the fact that the general formula for the probability distribution can be obtained by expanding a binomial expression with negative index. It can be shown that $P(X=r)$ is the term involving q^{r-k} in the binomial series expansion of $p^k(1-q)^{-k}$, for $r=k,\ k+1,\ k+2,\ \ldots$, the number of possible values being countably infinite.

As in the case of the geometric distribution, there is an alternative form in which X is defined as the number of *failures before* the kth success. This has the effect of moving the whole distribution down k places, so that $P(X=r)$ is the term involving q^r in the binomial series expansion of $p^k(1-q)^{-k}$, for $r=0$ or any positive integer.

Reverting to the original form of the distribution, which is the one tabulated by Program 6.5, the relationship of successive terms in the probability distribution is given by:

$$P(X=r+1) = \frac{r}{r+1-k}qP(X=r),\ r = k,\ k+1,\ k+2,\ \ldots$$

The mean and variance are given by:

$$\mu = \frac{k}{p} \qquad \sigma^2 = \frac{kq}{p^2}$$

The reader will of course wish to assure himself that all the above formulae, starting from the expansion of $p^k(1-q)^{-k}$, fit the results for the geometric distribution when k is 1.

The negative binomial distribution has some unexpected applications. It can be shown theoretically that under certain general assumptions about the differing 'accident-proneness' of different individuals, each individual having accidents purely at random but at a rate determined by his own particular accident-proneness, the overall statistics of the numbers of individuals who sustain 0, 1, 2, . . ., accidents in a fixed period will fit a negative binomial distribution of the 'alternative' form. The same considerations may apply to many forms of breakdown in science or engineering.

For the 'alternative' form, $\mu = kq/p$ while σ^2 remains at kq/p^2. We can re-arrange this as:

$$p = \frac{\mu}{\sigma^2} \qquad k = \frac{\mu^2}{\sigma^2 - \mu}$$

We would be entitled to suspect that there may be 'accident-proneness' if for the collected data of accidents (or, of course, breakdowns in some industrial process or construction method) the observed variance is larger than the observed mean. Taking these as estimates of σ^2 and μ respectively, we estimate p and k and find the appropriate set of probabilities for the negative binomial distribution. It does not matter that k will almost certainly not be an integer; the theory is still applicable and Program 6.5 will still work; it is however necessary to subtract k both from the values of r and from the value for the 'Mean' printed out, and it should then be found that the adjusted mean and the variance agree with those of the data.

The relative frequencies may then be compared with the theoretical probabilities, and if they fit well (preferably confirmed by a goodness-of-fit test as presented in Chapter 11) it may be concluded that there is evidence of 'accident-proneness'. Conversely, but equally important in practice, if the observed variance is not much larger than the observed mean it is not reasonable to assume that the individuals or installations which have the highest numbers of accidents are accident-prone — they are just unlucky! It would be important to collect the data on each individual or installation for as many years as possible, taken together; if it is the same individual who appears on each year's evidence taken separately to be 'just unlucky', the accumulated accident figures will show him to be accident-prone.

Program 6.5

```
5000 REM NEGATIVE BINOMIAL DISTRIBUTION
5100 PRINT: PRINT: PRINT "  NEGATIVE BINOMIAL DISTRIBUTION"
5110 PRINT: PRINT "X is the number of Bernoulli trials up"
5120 PRINT "to and including the kth success, each"
5130 PRINT "trial having probability p of success."
5134 REM PUTTING k = 1 PRODUCES THE GEOMETRIC DISTRIBUTION.
5135 REM THE DISTRIBUTION OF (X-k), THE NUMBER OF FAILURES
5136 REM BEFORE THE kTH SUCCESS, IS ALSO CALLED A NEGATIVE
5137 REM BINOMIAL DISTRIBUTION.  IF k IS LARGE AND p IS
```

```
5138 REM CLOSE TO 1, (X-k) HAS APPROXIMATELY A POISSON
5139 REM DISTRIBUTION WITH MEAN EQUAL TO k*(1-p)/p.
5140 PRINT: INPUT "Input k,p ";K,P
5150 Q = 1-P: M = K/P: V = M*Q/P
5200 PRINT: PRINT "Mean = ";M
5210 PRINT "Variance = ";V
5220 PRINT "Standard Deviation = ";SQR(V)
5230 IF K*LOG(Q)<-18 GOTO 5600
5300 W = P^K
5310 Y = W: RL = K: RU = K+4
5400 PRINT
5420 PRINT: PRINT " r     P(X=r)";TAB(21);"P(X<=r)"
5430 PRINT
5440 FOR I = RL TO RU
5450 IF I<10 THEN PRINT " ";
5460 PRINT ;I;"    ";W;TAB(20);Y
5480 W = W*I*Q/(I+1-K): Y = Y+W
5490 NEXT I
5500 PRINT: PRINT "Press C to continue."
5510 PRINT: PRINT "Press R to return."
5520 R$ = GET$
5530 IF R$="R" GOTO 5100
5540 RL = RU+1: RU = RU+12
5550 PRINT: PRINT: PRINT "  NEGATIVE BINOMIAL DISTRIBUTION"
5560 PRINT: PRINT "      k = ";K;"  p = ";P
5570 GOTO 5400
5600 PRINT: PRINT "This program will not handle such
     a large k when p = ";P
5610 GOTO 5100
```

Program notes

(1) For general comments on the style and layout of this program, see the notes to Programs 6.1 and 6.2.

(2) At line 5300, $P(X=k)$ is evaluated as the first term in the distribution. If this is too small for the computer, lines 5230 and 5600 come into effect. Note that LOG as used here means the logarithm to base 10.

(3) At line 5480, W is set at $P(X=I+1)$ using the relationship with $P(X=I)$ discussed in the text.

6.8 The Poisson distribution

The Poisson distribution is the distribution of the number of random happenings in a fixed interval of a continuum. The happenings must be *completely* random; that is, the probability that a happening will occur in the next second (or in the next year) is not influenced at all by how long it is since the last happening took place.

Good examples are cosmic rays entering a Geiger counter, alpha particles emitted by a radioactive mass, and thread breakages in textile processing. The continuum may be in space rather than time; the numbers of oilstains in 10-metre lengths of fabric, the numbers of yeast cells in the squares ruled on a glass slide in a haemocytometer, and the numbers of currants in the one-inch cubes of a currant cake in which the fruit did not sink, can all be expected to fit a Poisson distribution. Many textbooks quote examples where the property of randomness is highly suspect, such as the numbers of vehicles passing along a road.

The only parameter of the distribution is its mean, and it is convenient to put μ directly into the formula for the distribution, which is:

$$P(X=r) = \frac{e^{-\mu}\mu^r}{r!}, \quad r = 0, 1, 2, \ldots$$

As for most of the distributions in this chapter there is more than one way to derive this formula, but we shall not attempt to derive it here. When one remembers the power series expansion of e^{μ} it is immediately obvious that the terms in the distribution sum to 1, which is of course an essential property of every probability distribution.

The relationship of successive terms is given by:

$$P(X=r+1) = \frac{\mu}{r+1} P(X=r), \quad r = 0, 1, 2, \ldots$$

The method of obtaining the mean without recourse to probability generating functions is typical of the methods appropriate to several types of probability distribution, but simpler to perform than most:

$$\begin{aligned}\sum_{r=-\infty}^{\infty} rP(X=r) &= \sum_{r=1}^{\infty} r\frac{e^{-\mu}\mu^r}{r!} = \sum_{r=1}^{\infty} \frac{e^{-\mu}\mu^r}{(r-1)!} \\ &= \sum_{s=0}^{\infty} \frac{e^{-\mu}\mu^{s+1}}{s!} = \mu e^{-\mu}\sum_{s=0}^{\infty}\frac{\mu^s}{s!} = \mu e^{-\mu}e^{\mu} \\ &= \mu.\end{aligned}$$

The typical 'tricks' in this calculation were to note that we could eliminate the term involving $r=0$ from the summation (since it makes no contribution to the mean) before writing $(r-1)!$, and to make the substitution $s=r-1$ to obtain the power series expansion of e^{μ} which we could then sum.

To find the variance, it is simplest to show that

$$\sum_{r=-\infty}^{\infty} r(r-1)\, P(X = r)$$

is equal to μ^2 and hence that $\sigma^2 = \mu$. The fact that the variance of a Poisson distribution is equal to its mean justifies our assertion in the previous section that if the observed variance of the data on numbers of accidents is not much larger than the observed mean then it is not reasonable to assume any tendency to accident-proneness, since it means that the data are consistent with the hypothesis that all accidents are random happenings with the same value of μ for everyone.

It can be shown that a negative binomial distribution of the 'alternative form' tends to a Poisson distribution if k is large and p is close to 1 (which makes σ^2 close to μ), and so in these circumstances the Poisson distribution could be used as an approximation to the negative binomial distribution. It can similarly be shown that a binomial distribution tends to a Poisson distribution if n is large and p is small (which again makes σ^2 close to μ, though this time it is less than μ). The Poisson distribution is therefore widely used as an approximation to the binomial distribution when p is less than about 0.1 and n is at least 20. There is not much point in using these approximations when a computer is available and can produce the exact distributions quite readily, but combining the distributions in a computer package offers a good opportunity to examine how close the approximations actually are.

We use the word 'happenings' in the definition of a Poisson variable since the term 'events' leads to confusion with the technical definition of 'event' as explained in Section 4.2. If X has a Poisson distribution then $P(X=0)$ is an example of the probability of an event.

Further confusion arises from texts which imply that the Poisson distribution is concerned with 'the occurrence of rare events' (S. E. Hodge and M. L. Seed (1972)). If cosmic rays occur completely at random, the number occurring in a fixed interval of time has a Poisson distribution irrespective of whether the interval is 1 millisecond or 1 hour; the respective means are in proportion to the lengths of the intervals. The same principles apply to random events whose mean rate is 1 per millennium. The additive property of Poisson variables for neighbouring (but not for overlapping) intervals is one of their essential characteristics.

In Program 6.6 the mean of a Poisson distribution is limited to 43, but this is only because $P(X=0)$ is otherwise too small for the BASIC to cope with.

Program 6.6

```
6000 REM POISSON DISTRIBUTION
6100 PRINT: PRINT: PRINT "     POISSON DISTRIBUTION"
6110 PRINT: PRINT "X is the number of random happenings in"
6120 PRINT "a fixed interval of a continuum, where"
6130 PRINT "the mean number of happenings for that"
6140 PRINT "size of interval is known."
6150 PRINT: INPUT "Input mean ";M
6200 PRINT: PRINT "Mean = ";M
6210 PRINT "Variance = ";M
6220 PRINT "Standard Deviation = ";SQR(M)
6230 IF M>43 GOTO 6600
6300 W = EXP(-M)
6310 Y = W: RL = 0: RU = 4
6400 PRINT
6420 PRINT: PRINT " r     P(X=r)";TAB(21);"P(X<=r)"
6430 PRINT
6440 FOR I = RL TO RU
```

```
6450 IF I<10 THEN PRINT " ";
6460 PRINT ;I;"     ";W;TAB(20);Y
6480 W = W*M/(I+1): Y = Y+W
6490 NEXT I
6500 PRINT: PRINT "Press C to continue."
6510 PRINT: PRINT "Press R to return."
6520 R$ = GET$
6530 IF R$="R" GOTO 6100
6540 RL = RU+1: RU = RU+12
6550 PRINT: PRINT: PRINT "     POISSON DISTRIBUTION"
6560 PRINT: PRINT "         Mean = ";M
6570 GOTO 6400
6600 PRINT: PRINT "This program will not handle so large
     a mean."
6610 GOTO 6100
```

Program notes

For general notes on the style and layout of this program, see the notes to Programs 6.1 and 6.2. If the mean is greater than 43, line 6230 rejects it rather than produce a DATA ERROR at line 6300. $P(X = 0)$ is evaluated at line 6300, and at line 6480 further values are calculated from the relationships between $P(X = r+1)$ and $P(X = r)$ presented in the text.

6.9 The uniform discrete distribution

One of the simplest probability distributions is that of a discrete random variable which is equally likely to be any of K successive integers. An obvious example is the number thrown on an unbiased die, when $K = 6$.

This uniform discrete distribution is defined by two parameters. One is the lowest integer which the random variable can take, which we denote by L. The other may be either the highest integer which the random variable can take, denoted by H, or — as we have preferred — the number of possible values K. Clearly $H = L+K-1$. The probability distribution is:

$$P(X = r) = \frac{1}{K}, \ r = L, L+1, L+2, \ldots, L+K-1$$

The mean is the mid-range, $(L+H)/2$, which is equal to $L+(K-1)/2$. The variance is readily computed using the fact that the sum of the squares of the first n integers is $n(n+1)(2n+1)/6$ and assuming that L is a positive integer. It turns out that $\sigma^2 = (K^2-1)/12$, and that this remains true when L is negative.

If $L = 0$ and $K = 2$, the distribution becomes a Bernoulli distribution with $p = ½$, and Programs 6.1 and 6.7 both give $\mu = ½, \sigma^2 = ¼$.

Program 6.7

```
7000 REM UNIFORM DISCRETE DISTRIBUTION
7100 PRINT: PRINT: PRINT "  UNIFORM DISCRETE DISTRIBUTION"
7110 PRINT: PRINT "X is a discrete random variable which"
7120 PRINT "has uniform probability 1/K of taking"
```

```
7130 PRINT "any integral value from L to L+K-1."
7140 PRINT: PRINT "Input lowest attainable value (L),"
7150 INPUT "number of attainable values (K) ";L,K
7160 M = L+(K-1)/2: V = (K*K-1)/12
7200 PRINT: PRINT "Mean = ";M
7210 PRINT "Variance = ";V
7220 PRINT "Standard Deviation = ";SQR(V)
7300 W = 1/K
7310 Y = W: T = L+K-1
7320 RL = L: RU = L+4: IF T<RU THEN RU = T
7400 PRINT
7420 PRINT: PRINT " r      P(X=r)";TAB(21);"P(X<=r)"
7430 PRINT
7440 FOR I = RL TO RU
7450 IF I<10 THEN PRINT " ";
7460 PRINT ;I;"    ";W;TAB(20);Y
7480 Y = Y+W
7490 NEXT I
7500 IF RU<T THEN PRINT: PRINT "Press C to continue."
7510 PRINT: PRINT "Press R to return."
7520 R$ = GET$
7530 IF R$="R" OR RU=T GOTO 7100
7540 RL = RU+1: RU = RU+12: IF T<RU THEN RU = T
7550 PRINT: PRINT: PRINT "  UNIFORM DISCRETE DISTRIBUTION"
7560 PRINT: PRINT "     Range ";L;" to ";T
7570 GOTO 7400
```

Program notes

For general notes on the style and layout of this program, see the notes to Programs 6.1 and 6.2.

PROBLEMS

(6.1) Apply Program 6.1 to tabulate the probabilities for the Bernoulli distribution for $p = 0.05, 0.1, 0.5$, and 0.9. Check these results using the binomial distribution Program 6.2, and find the probability distributions for the same values of p for $n = 2, 3, 4, 5$. Confirm that the distribution for $p = 0.5$ is symmetrical, and that the distribution for $p = 0.9$ is the reverse of that for $p = 0.1$. Then find the probability distributions using Program 6.4 for the hypergeometric distribution with the same values of n, putting $N = 20$ and $S = 1, 2, 10$ and 18, and compare each result with the bionominal distribution equivalent when $p = S/N$.

(6.2) Apply Program 6.2 to tabulate all probabilities greater than 0.00005 when $n = 20$ and $p = 0.025, 0.05$, and 0.1. Then apply Program 6.6 to tabulate all probabilities greater than 0.00005 when $\mu = 0.5, 1$, and 2, and compare the results with the binomial distributions for the same values of $\mu = np$.

(6.3) Compare the negative binomial distribution with the geometric distribution and with the Poisson distribution in an analogous manner to the comparisons applied to the binominal distribution in problems 6.1 and 6.2, applying the information in lines 5135–5139 of Program 6.5.

(6.4) Develop Program 6.2 to offer the option of obtaining a single probability, e.g. $n = 50, p = 0.4, r = 25$, without having to list all the probabilities for lower values of r. Similarly develop all the Programs 6.3 to 6.7.

(6.5) Find the ways in which Programs 6.1 to 6.7 will fail if incorrect values are inputted, such as negative values for n or values of p greater than 1. Develop each program so that if an impossible value is inputted the user is told what is wrong and invited to input a genuine value. Note however that, for instance, the negative binomial distribution can be applied with k taking positive non-integer values.

(6.6) Try to develop Program 6.2 so that it will calculate probabilities for values close to the mean even when the probabilities of zero successes and of n successes are both too small for the computer to handle. (Hint: *either* start by calculating the largest probability, for the integer closest to the mean, and then work downwards as well as upwards to cover all values with non-negligible probabilities, *or* start at the integer about three standard deviations below the mean and work upwards. Each of these methods has its disadvantages.)

(6.7) Program 6.4 fails when the probabilities of zero successes and of n successes are both zero. This means that it cannot cope even with simple examples such as $n = 3, N = 4, S = 2$. Develop the program so that it will work for all values of n, N, and S as long as the probabilities are not extremely small.

(6.8) Find the mean and variance of the following data, fit a negative binomial distribution as discussed in section 6.7, and compare the expected frequencies (probabilities multiplied by 100) with the observed frequencies. Keep the results for a goodness-of-fit test in chapter 11.

No. of accidents in the year:	0	1	2	3	4	5	6	7
No. of cases:	25	25	17	12	8	7	4	2

(6.9) The 'horse-kick' Data 3 in section 2.5 was considered suitable for comparison with a Poisson distribution having the same mean, $\mu = 0.7$. Use Program 6.6 to find this distribution, and compare the expected frequencies (probabilities multiplied by 280) with the observed frequencies. Keep the results for a goodness-of-fit test in chapter 11.

(6.10) Confirm that the same probability distribution and the same mean and variance are produced by Programs 6.1 with $p = ½$, 6.2 with $n = 1, p = ½$, 6.4 with $n = 1, N =$ any even number, $S = N/2$, and 6.7 with $L = 0, K = 2$.

Chapter 7

Continuous random variables and probability density functions

7.1 The properties of continuous random variables

Weights in kilogrammes, heights in centimetres, etc., are examples of continuous data; they can take any value — integral, fractional, or irrational — within their range. The sample space for the experiment of selecting a male student at random and measuring his height in centimetres (that is, taking 1 cm as the unit but not rounding the measurement to the nearest integral number of centimetres) is a *continuous sample space*; it contains an infinite number of sample points in any finite interval of the space.

Every sample point in a continuous sample space has probability zero, but we can assign finite probabilities to finite intervals. For instance, we might well find in the foregoing experiment that the probability of a value between 174.5 and 175.5 (using the limiting relative frequency definition of probability) is 0.05. Since the total probability for the whole sample space is 1, it must follow that the probabilities for values below 174.5 and above 175.5 would sum to 0.95.

A *continuous random variable* is a random variable defined on a continuous sample space. Almost invariably the relationship is the simple straightforward relationship which in the case of discrete random variables we saw only in the case of a throw of a die; the random variable is simply the value observed. In the foregoing experiment most statisticians would amalgamate the description of the experiment, the delineation of the sample space, and the definition of the random variable into a single phrase: 'let X be the height in centimetres of a randomly-selected male student'.

For this experiment there is unlikely to be any appeal to theory in order to determine the probabilities. On the basis of repeated observations or measurements we would obtain some estimate of the probability that X lies between 174.5 and 175.5, or any other interval in which we are interested. It would not however be necessary to list all conceivable intervals, but only to estimate the form of the cumulative distribution function, which we defined in Chapter 6 as:

$$F(x) = P(X \leqslant x).$$

Then $\quad P(174.5 < X < 175.5) = F(175.5) - F(174.5),$

since the sample point 175.5 has zero probability.

In general, for both discrete and continuous random variables, $F(x)$ must be a non-decreasing function of x which tends to zero as x tends to $-\infty$ and to 1 as x tends to ∞.

It is usually convenient also to know the *probability density function*, $f(x)$, which is the derivative of $F(x)$ with respect to x:

$$f(x) = \frac{\mathrm{d}}{\mathrm{d}x}F(x) \text{ and hence } f(x) \geqslant 0 \text{ for all } x.$$

$$F(x) = \int_{-\infty}^{x} f(u)\,\mathrm{d}u \text{ and hence } P(a < X \leqslant b) = F(b) - F(a)$$

$$= \int_{a}^{b} f(x)\,\mathrm{d}x, \text{ provided that } a < b.$$

The mean and variance are then defined in an analogous way to the definitions for discrete random variables:

$$\mu = \int_{-\infty}^{\infty} xf(x)\,\mathrm{d}x \qquad \sigma^2 = \int_{-\infty}^{\infty} (x-\mu)^2 f(x)\,\mathrm{d}x$$

and it is not difficult to obtain an identity analogous to that presented in Section 6.2 and equally useful:

$$\sigma^2 = \mu'_2 - \mu^2 = \int_{-\infty}^{\infty} x^2 f(x)\,\mathrm{d}x - \mu^2$$

7.2 The negative exponential distribution

In Section 6.8 it was stated that if cosmic rays enter a Geiger counter completely at random, then the number of such happenings in a fixed interval of time will have a Poisson distribution. Since Poisson variables for neighbouring intervals are additive, it follows that if the mean rate per second is λ then the mean number of happenings for a fixed interval of x seconds is (λx).

By using the probability of a Poisson variable taking the value zero, it can be seen that the probability of no cosmic rays in time x is $e^{-\lambda x}$. So if we define the continuous random variable X as the time in seconds until the first cosmic ray enters a Geiger counter when the mean rate per second is λ, the Law of Complementarity leads immediately to the result:

$$F(x) = 1 - e^{-\lambda x}, \; x \geqslant 0.$$

This is the *negative exponential distribution*. Differentiating $F(x)$ with respect to x gives:

$$f(x) = \lambda e^{-\lambda x}, \quad x \geqslant 0.$$

By integration by parts (or by making use of the formulae for the Laplace transforms of x and x^2), it can be shown that:

$$\mu = \int_0^\infty x\lambda e^{-\lambda x} dx = \frac{1}{\lambda}$$

$$\sigma^2 = \int_0^\infty x^2 \lambda e^{-\lambda x} dx - \left(\frac{1}{\lambda}\right)^2 = \frac{2}{\lambda^2} - \frac{1}{\lambda^2} = \frac{1}{\lambda^2}$$

So the standard deviation is equal to the mean. The moments about the origin follow the formula

$$\mu'_r = \frac{r!}{\lambda^r}$$

as may be seen most easily using Laplace transforms. So the moments about the mean are

$$\mu_3 = \frac{2}{\lambda^3} \quad \text{and} \quad \mu_4 = \frac{9}{\lambda^4}$$

and the coefficients of skewness and kurtosis are therefore 2 and 9 respectively. These constitute a strong positive skew and an extremely high kurtosis.

The negative exponential distribution is sometimes known simply as 'the exponential distribution'.

There is nothing special about the role of the constant 'e' in the negative exponential distribution. What we are concerned with is a strictly non-negative continuous random variable whose probability density function is proportional to a^x, where a lies between 0 and 1. It will be found that in order that the total probability should be 1 the coefficient of a^x must be $\log_e(1/a)$, and that the mean is then $1/\log_e(1/a)$. Both formulae are made to look much simpler by writing $\lambda = \log_e(1/a)$, and this makes a equal to $e^{-\lambda}$.

Writing the cumulative distribution function in the form $1-a^x$ emphasizes the relationship between the negative exponential distribution and the geometric distribution, with a and q as respective parameters and x and r as the respective values which the random variables take. The negative exponential distribution is the limiting form of the geometric distribution, while the Poisson distribution is the limiting form of the binomial distribution, as a sequence of independent trials tends to a continuum in which there are random happenings.

7.3 The normal or Gaussian distribution

Just as random happenings give rise to a probability density function proportional to a to the power x, so random errors give rise to a probability density function proportional to a to the power x^2. Once again the parameter a must lie between 0 and 1 in order that the area under the curve can be finite, but we now have the reasonable assumption that errors (unlike the intervals between random happenings) are equally likely to be negative instead of positive.

The curve $f(x) = k a^{x^2}$, where k is a constant and a lies between 0 and 1, is a symmetric 'bell-shaped' curve which slopes downwards on both sides from $x = 0$. It takes the value k at $x = 0$, and for large positive x or numerically-large negative x it approaches the x-axis very closely but never touches it. The mean is of course zero, and the standard deviation depends on the value of a; the smaller the value of a the more rapidly the curve falls away and hence the smaller the standard deviation, the relationship in fact being found to be:

$$\sigma^2 = 1/(2 \log_e (1/a))$$

Once again we have a case where it is simpler to express $f(x)$ in terms of something other than a, and in fact it is customary to put both a and k in terms of σ. The distribution can also be made more general by replacing x by $(x - \mu)$ so that it is the distribution of actual observations rather than simply of 'errors'. The general form of the 'normal' or 'Gaussian' distribution is then:

$$f(x) = \frac{1}{\sigma \sqrt{2\pi}} \exp \left[\frac{-(x-\mu)^2}{2\sigma^2} \right]$$

Presented in this way the distribution looks formidable at first sight. Note carefully that it has been deliberately arranged in terms of its parameters μ and σ, and that the irrational constants e and π then arise from the mathematical relationships between the probability density function and the parameters.

The mathematical justification of the normal distribution is that it is the limiting form of the distribution of the sum of a large number of independent random variables of which no one variable 'dominates' the others. Since a large number of independent causes contribute to the size of a ball bearing, the diameter of a yarn, the crushing load of a concrete cube, etc., it is not surprising that all these types of manufacturing measurement are found to conform very closely to normal distributions, each of course with its own typical values for μ and σ. It is also found that natural measurements, such as the heights and weights of adult males from a homogeneous population, have normal distributions.

We have now expounded basic ideas underlying the normal distribution which are often omitted from statistics textbooks, and can safely leave the reader to his statistics textbook for the usual diagrams and the explanations of how the normal distribution is 'applied'. But before turning to the computing aspects, we give three warnings against possible misconceptions and hence misuse.

Firstly, the impression is sometimes given that the normal distribution is applicable to virtually every kind of measurement of a continuous variable. As already indicated, measurements of 'running times' of machines subject to random stoppages, and in general of intervals between random happenings, will have a negative exponential distribution — which can never remotely resemble a normal distribution.

Secondly, the discerning reader will have noted the oddity of claiming that measurements of natural or manufactured phenomena, which clearly must be positive, can be fitted to a type of distribution which allots positive probabilities to negative values. The sounder claim is that it is the *logarithms of the measurements* which have normal distributions. If the mean of the measurements is at least four times the standard deviation, the error in assuming that the measurements themselves have approximate normal distributions is negligible. But if this is not so, as for instance in measuring the diameters or volumes of particles of powdered detergent (which have distributions with so strong a skew to the right that the mode takes a value not much larger than the standard deviation), then it is quite essential to take the logarithms of the measurements as the data to which any statistical technique which assumes an approximate normal distribution is applied.

Thirdly, it is often stated (quite correctly) that the sum or mean of a number of independent random variables with identical distributions will have an approximate normal distribution if the number is large enough. This is the Central Limit Theorem. But even the best texts sometimes assert quite falsely that the approximation will be good if the number (i.e. the sample size, assuming that the random variables are measurements on specimens drawn from the same population) is at least 30 'regardless of the shape of the population' (Ronald E. Walpole (1982)). In fact, if the sample is drawn from a Bernoulli distribution with $p=0.001$ or $p=0.999$ it will be necessary for the sample size to be at least 4000 before approximate normality is attained for the sample mean.

To find the probability that X lies in the interval between a and b we need to know the integral of $f(x)$ between a and b. Unfortunately there is no explicit formula for $F(x)$ — in other words, $f(x)$ cannot be integrated analytically. We make use of the fact that:

$$P(a < X \leqslant b) = \int_a^b \frac{1}{\sigma\sqrt{2\pi}} \exp\left[\frac{-(x-\mu)^2}{2\sigma^2}\right] dx$$

$$= \int_{z_a}^{z_b} \frac{1}{\sqrt{2\pi}} \exp\left(\frac{-z^2}{2}\right) dz$$

where $z_a = \dfrac{a-\mu}{\sigma}$ and $z_b = \dfrac{b-\mu}{\sigma}$, and assuming that $a < b$.

The relationship is easily obtained using the substitution

$$z = \frac{x-\mu}{\sigma}$$

and means that all problems involving normal distributions with known μ and σ can be solved using the table of the *standard normal distribution:*

$$\Phi(x) = \int_{-\infty}^{x} \frac{1}{\sqrt{2\pi}} \exp\left(\frac{-z^2}{2}\right) dz$$

For instance, if we know that the height in centimetres of a randomly-selected male student has mean 175 and standard deviation 8, we will certainly assume that the distribution is normal and therefore conclude that the probability of an observation between 174.5 and 175.5 is:

$$F(175.5)-F(174.5) = \Phi\left(\frac{175.5-175}{8}\right) - \Phi\left(\frac{174.5-175}{8}\right)$$

$$= \Phi(0.0625) - \Phi(-0.0625) = 0.5249-0.4751$$

$$= 0.0498.$$

In Program 7.1, $F(x)$ can be found for any values of μ, σ, and x. The value of the standard normal variable z is obtained as above, and the algorithm then used to find $F(x) = \Phi(z)$ is based on the fact that $\Phi(0) = 0.5$ together with the series expansion obtainable by repeated integration by parts:

$$\int_0^x \exp\left(\frac{-z^2}{2}\right) dz = \exp\left(\frac{-x^2}{2}\right)\left(x + \frac{x^3}{3} + \frac{x^5}{3*5} + \frac{x^7}{3*5*7} + \ldots\right)$$

This algorithm is slow compared with some of the fast algorithms which have been developed, and is not to be recommended for use in a simulation loop. It is also very slow if x is large. However, it is found that for x greater than 5.6, $\Phi(x) = 1$ to about 8 places of decimals, and similarly we can put $\Phi(x)=0$ for $x< -5.6$. With the extreme values taken care of in this way, the algorithm is quite fast enough for most purposes and is certainly the simplest algorithm to understand and to program.

Program 7.1 includes five options, but for the moment we shall look only at the first two options, that is, the lines up to line 2300 ignoring the references to the variable *WN* in lines 570–580. An important feature of the program is that if it is used to evaluate $F(x)$ for two values of x without changing the parameters of the distribution it prints out the difference between the results and hence the probability for the interval between the two values of x.

Program 7.1

```
100 REM CONTINUOUS RANDOM VARIABLES
110 EP = 1E-9: C1 = 5.6: C2 = 1/SQR(2*PI): C3 = 1/SQR(12)
120 PRINT: PRINT: PRINT " 1  NEGATIVE EXPONENTIAL DISTRIBUTION"
130 PRINT "    (also called EXPONENTIAL DISTN.)"
140 PRINT: PRINT " 2  NORMAL DISTRIBUTION"
150 PRINT "    (also called GAUSSIAN DISTN.)"
160 PRINT: PRINT " 3  NORMAL DISTRIBUTION used as an"
170 PRINT "    approximation to the BINOMIAL DISTN."
180 PRINT: PRINT " 4  NORMAL DISTRIBUTION used as an"
190 PRINT "    approximation to the POISSON DISTN."
200 PRINT: PRINT " 5  RECTANGULAR DISTRIBUTION"
210 PRINT "    (also called UNIFORM DISTN.)"
220 PRINT: PRINT: PRINT "  Which kind of distribution do"
230 PRINT "  you want to study?"
240 PRINT: PRINT "  Press 1, 2, 3, 4, or 5."
250 D = VAL(GET$): WN = 4: R$ = "R"
260 ON D GOTO 1000,2000,3000,4000,5000
400 PRINT: PRINT "Mean = ";M
410 PRINT "Variance = ";V
420 PRINT "Standard Deviation = ";S
430 PRINT: PRINT "F(x) is the probability that X<=x."
440 PRINT: INPUT "Input the value x at which F(x) is to be
    evaluated ";X
450 RETURN
500 IF Z<-C1 THEN Y = 0: GOTO 570
510 IF Z>C1 THEN Y = 1: GOTO 570
520 ZZ = Z*Z: K = C2*EXP(-ZZ/2)
530 IF Z<0 THEN Z = -Z: K = -K
540 Y = Z: I = 1
550 I = I+2: Z = Z*ZZ/I: Y = Y+Z: IF Z>EP GOTO 550
560 Y = 0.5+K*Y
570 PRINT: PRINT "F(";X;") = ";Y;: IF WN<2 THEN PRINT
    " NOTE EMPHATIC";
580 PRINT: IF WN<3 THEN PRINT "WARNING: Normal approx.
    is not reliable!"
590 IF R$="N" THEN PRINT: PRINT "F(";X;")-F(";T;") = ";Y-U
600 T = X: U = Y
610 PRINT: PRINT "Press N to input a new value of x."
620 PRINT: PRINT "Press R to return."
630 R$ = GET$
640 IF R$="R" GOTO 120
650 PRINT: INPUT "Input the value x at which F(x) is to be
    evaluated ";X
660 ON D GOTO 1300,2300,3300,4300,5300
1000 REM NEGATIVE EXPONENTIAL DISTRIBUTION
1100 PRINT: PRINT: PRINT "   NEGATIVE EXPONENTIAL DISTRIBUTION"
1110 PRINT: PRINT " (Also called Exponential Distribution)"
1120 PRINT: PRINT "X has a negative exponential distribu-"
```

```
1130 PRINT "tion with parameter L, and hence with"
1140 PRINT "mean 1/L."
1200 PRINT: INPUT "Input L ";L
1210 M = 1/L: S = M: V = S*S: GOSUB 400
1300 Y = 1-EXP(-L*X): GOTO 570
2000 REM NORMAL DISTRIBUTION
2100 PRINT: PRINT: PRINT "    NORMAL DISTRIBUTION"
2110 PRINT: PRINT " (Also called Gaussian Distribution)"
2120 PRINT: PRINT "Mean m, standard deviation s."
2200 PRINT: INPUT "Input m,s ";M,S
2210 V = S*S: GOSUB 400
2300 Z = (X-M)/S: GOTO 500
3000 REM BINOMIAL DISTRIBUTION
3100 PRINT: PRINT: PRINT "BINOMIAL DISTRIBUTION (NORMAL APPROXN.)"
3110 PRINT: PRINT "n trials, probability p of success."
3200 PRINT: INPUT "Input n,p ";N,P
3210 M = N*P: V = M*(1-P): S = SQR(V): GOSUB 400
3220 WN = M/S: IF N-M<M THEN WN = (N-M)/S
3300 Z = (INT(X)+0.5-M)/S: GOTO 500
4000 REM POISSON DISTRIBUTION
4100 PRINT: PRINT: PRINT "POISSON DISTRIBUTION (NORMAL APPROXN.)"
4200 PRINT: INPUT "Input mean ";M
4210 S = SQR(M): V = M: GOSUB 400
4220 WN = S
4300 Z = (INT(X)+0.5-M)/S: GOTO 500
5000 REM RECTANGULAR DISTRIBUTION
5100 PRINT: PRINT: PRINT "    RECTANGULAR DISTRIBUTION"
5110 PRINT: PRINT " (Also called Uniform Distribution)"
5120 PRINT: PRINT "X lies between A and B with uniform probability."
5200 PRINT: INPUT "Input A,B ";A,B
5210 M = (A+B)/2: S = C3*(B-A): V = S*S: GOSUB 400
5220 K = 1/(B-A)
5300 Y = K*(X-A): IF Y<0 THEN Y = 0
5310 IF Y>1 THEN Y = 1
5320 GOTO 570
9999 END
```

Program notes

(1) The values of EP and C1 are used at lines 500–550. If 7-figure accuracy is not required, EP can be larger and C1 correspondingly smaller (see Problems). Some computers offer PI or π on call; if not, put PI as 3.1415927. Better still, whether PI is available or not, is to put C2 = 0.39894228 which is correct to 9 significant figures since the 9th figure is zero.

(2) C3 at line 110 is not required for the first two options, and WN = 4 at line 250 renders the conditional statements at lines 570–580 inoperative.

(3) If you have trouble with lines 250–260, see note (2) to Program 5.3.

(4) Lines 400– 450 form a simple subroutine which is called at lines 1210 and 2210 when the mean and variance are known.

(5) Lines 500–560 compute $\Phi(z)$ as described in the text.

(6) Lines 570–650 print out $F(x)$, and if R$ = "N" the difference from the previous value of $F(x)$ — which involved the same parameters — is also printed out.

(7) For line 660 see note (3) above.

(8) Lines 1000–1300 produce $Y = F(x)$ for the negative exponential distribution, using $L = \lambda$, $M = \mu$, $S = \sigma$.

(9) Lines 2000–2300 produce Z for the normal distribution and then go to the subroutine at line 500 to compute $F(x) = \Phi(z)$.

7.4 Normal approximations to discrete distributions

It has already been indicated that the sum or mean of a number of independent random variables with identical distributions will have an approximate normal distribution if the number is large enough. It follows that any distribution which is *itself* the distribution of the sum of independent identically-distributed random variables must have the form of an approximate normal distribution for suitable values of the parameters. The binomial distribution and the negative binomial distribution obviously qualify for consideration, though the latter is less important. When one remembers the additive property of independent Poisson variables, so that any Poisson variable with mean μ can be regarded as the sum of n independent Poisson variables each with mean μ/n, it is clear that the Poisson distribution must also have approximately the form of a normal distribution if μ is large enough.

An apparent difficulty arises from the fact that for discrete distributions only single points have positive probabilities, while for the normal distribution the probabilities are assigned to intervals. The solution is simple; if a discrete random variable X has a distribution which approximates to that of a normally-distributed random variable Y, then:

$$\begin{aligned} P(X \leqslant r) &= P(X < r+1) \simeq P(Y \leqslant r+\tfrac{1}{2}) = F_Y(r+\tfrac{1}{2}) \\ P(X \geqslant r) &= P(X > r-1) \simeq P(Y > r-\tfrac{1}{2}) = 1-F_Y(r-\tfrac{1}{2}) \\ P(X = r) &\simeq P(Y \leqslant r+\tfrac{1}{2}) - P(Y \leqslant r-\tfrac{1}{2}) = F_Y(r+\tfrac{1}{2})-F_Y(r-\tfrac{1}{2}) \end{aligned}$$

provided of course that (as in the case of all the discrete random variables we have considered) X takes the values of *integers*, r is an integer, and $F_Y(y)$ is the cumulative distribution function of Y.

Consider the binomial distribution with parameters 4, ½, and

r	*Binomial distribution* $P(X=r)$	$F(r)$	*Normal distribution* $F(r-\frac{1}{2})$	$F(r+\frac{1}{2})-F(r-\frac{1}{2})$
0	0.0625	0.0625	0.0062	0.0606
1	0.2500	0.3125	0.0668	0.2417
2	0.3750	0.6875	0.3085	0.3830
3	0.2500	0.9375	0.6915	0.2417
4	0.0625	1.0000	0.9332	0.0606
5	0	1.0000	0.9938	

hence with $\mu = np = 2$ and $\sigma = \sqrt{(npq)} = 1$. We can compare this distribution with the normal distribution with the same values of μ and σ.

It will be seen that the error in the approximation is never more than 0.0083 in $P(X=r)$ and never more than 0.0062 if we take $F(r+\frac{1}{2})$ for the normal distribution as an approximation to $F(r)$ for the binomial distribution. We usually regard errors of not more than 0.01 as constituting an acceptable approximation, though not necessarily a 'good' approximation.

If we consider the binomial distribution with parameters 100, 0.01, and hence $\mu = 1$, $\sigma = 0.9950$, the first few results are:

	Binomial distribution		*Normal distribution*	
r	$P(X=r)$	$F(r)$	$F(r-\frac{1}{2})$	$F(r+\frac{1}{2})-F(r-\frac{1}{2})$
0	0.3660	0.3660	0.0658	0.2419
1	0.3697	0.7357	0.3077	0.3846
2	0.1849	0.9206	0.6923	0.2419
3	0.0610	0.9816	0.9342	0.0598
4	0.0149	0.9965	0.9940	0.0058
5	0.0029	0.9994	0.9998	0.0002
6	0.0005	0.9999	1.0000	0.0000

$P(X=r)$ for the $b(100,0.01)$ distribution is the probability that the mean of 100 Bernoulli random variables with parameter 0.01 will take the value $0.01r$. Clearly the approximation is extremely poor, with errors of over 0.05 in both $P(X=r)$ and the cumulative probabilities; so much for those who claim that the approximation will be good for samples of size 30 or more 'regardless of the shape of the population'!

A condition for the approximation to be good is that $F(-\frac{1}{2})$ and $F(n+\frac{1}{2})$ for the normal distribution should be close to 0 and close to 1 respectively. Clearly the former of these provisos is not fulfilled for the $b(100,0.01)$ distribution, with the consequence that the normal approximation allots a probability of 0.0598 to the event that there are -1 successes! The alternative proviso would fail to precisely the same extent and with corresponding consequences in the case of the $b(100,0.99)$ distribution.

This leads to the suggested rule that we must have $2\sigma \leqslant \mu \leqslant n-2\sigma$ for the approximation to be given even the remotest consideration; we replace the figure 2 by 3 for a fairly good approximation and by 4 for a definitely good approximation to be hoped for. Even when a definitely good approximation is obtained, it is the *absolute* errors in probabilities which will be small. The proportionate errors in the case of small probabilities may still be very large, and of course there will be an infinite percentage overestimate of $P(X=-1)$ and

$P(X = n + 1)$ whose true probabilities are zero.

The real benefits of the normal approximation are in determining the probability for a large range of values of X. For instance, the $b(400,0.1)$ distribution has $\mu = 40$, $\sigma = 6$. $P(35 < X \leqslant 50)$ which is $P(36 \leqslant X \leqslant 50)$ can be obtained from the corresponding normal distribution as $F(50.5) - F(35.5) = 0.9599 - 0.2266 = 0.7333$. It would take a long time to calculate these probabilities correctly for a binomial distribution, and in fact Program 6.2 finds n too large to handle when $p = 0.1$ and $n \geqslant 394$.

All the considerations governing the use of the normal distribution as an approximation to the Poisson distribution have already been mentioned in relation to the binomial distribution. Since $\sigma = \sqrt{\mu}$ it follows that we can begin to consider the normal approximation when $\mu \geqslant 4$, expect a fairly good approximation when $\mu \geqslant 9$, and expect a good approximation when $\mu \geqslant 16$.

Program notes (for Program 7.1, lines 3000–4300)

(1) Line 3220 sets WN at whichever is the smaller of μ/σ or $(n - \mu)/\sigma$. Lines 570–580 then provide for a WARNING if WN is less than 3, which becomes an EMPHATIC WARNING if WN is less than 2. For more stringent conditions, 2 may be increased to 3 or even 4 at line 570, and 3 may be increased to 4 or even 5 or 6 at line 580; however, WN must be set at a value at line 250 which will not invoke the warning.

(2) Line 4220 sets WN equal to σ, which is the same as μ/σ for a Poisson distribution and so is appropriate for examining whether the conditions for an acceptable approximation are fulfilled.

(3) Lines 3210–3300 and 4210–4300 are designed to maintain the logic of two separate programs, in case it is desired to separate them and copy lines 400–440 and 500–650 into each of them. For the integrated program it would be a little more efficient to remove GOSUB 400 from line 3210, change line 3300 to GOTO 400, and put line 450 as IF D=3 OR D=4 THEN Z=(X+0.5−M)/S and line 660 as GOTO 450; similar changes would be needed to deal with the other options.

7.5 The rectangular or uniform distribution

A continuous random variable is said to have a 'rectangular' or 'uniform' distribution if the probability density function is a constant throughout the range from a to b and zero elsewhere. Since the total area under $f(x)$ must be 1, it follows that the constant must be $1/(b - a)$. The cumulative distribution function is then:

$$F(x) = \begin{cases} 0, & x < a \\ \dfrac{x-a}{b-a}, & a \leqslant x \leqslant b \\ 1, & x > b \end{cases}$$

Carrying out the integrations as indicated in Section 7.1 produces the results:

$$\mu = \frac{a+b}{2} \qquad \sigma^2 = \frac{(b-a)^2}{12}$$

The rectangular distribution is more often discussed for its mathematical interest than for its practical relevance. But it would arise if we were presented with observations which had been rounded to the nearest multiple of (say) 5 and wished to draw inferences from the true underlying values. We would then treat an observation such as 165 as representing a variable which could lie between 162.5 and 167.5 with uniform probability given by $f(x)=0.2$, $162.5 \leqslant x \leqslant 167.5$.

Program notes (for Program 7.1, lines 5000–5320)

(1) At line 5210 the value of σ is set at $(b-a)/\sqrt{12}$ using the constant C3 evaluated at line 110; there is no point in repeatedly employing the SQR function if it is not necessary. C3 could have been set directly at 0.28867513.

(2) At line 5300, $F(x)$ is set at $(x-a)/(b-a)$ whatever the value of x, and it is then corrected for $x < a$ or $x > b$. This is simpler than trying to put 'if $a \leqslant x \leqslant b$' into a statement.

7.6 Some other continuous random variables

The sum of a number of independent identically-distributed random variables each of which has a rectangular distribution tends to the form of a normal distribution as the number becomes large. The sum of 12 independent variables each of which is uniformly distributed between 0 and 1 (and hence has a square distribution) has a distribution which is sufficiently close to a normal distribution with mean 6 and variance 1 for it to be used as a way of simulating a normally-distributed random variable.

The sum of a number of independent random variables each of which has a normal distribution will itself have a normal distribution, even if the number is only 2 and even if the two means and variances differ from each other.

Sums of independent identically-distributed random variables each of which has a negative exponential distribution form a family of distributions known as the *gamma distribution,* the study of which is outside the scope of this text.

The distribution of the sum of the squares of ν independent standard normal variables (remembering that *standard* normal indicates that the mean is 0 and the variance is 1) is known as the *chi-square distribution with* ν *degrees of freedom*. This distribution, and the t-distribution and F-distribution which are derived from it, have very important applications in statistical inference but rather formidable probability density functions.

PROBLEMS

(7.1) The algorithm in lines 500–560 of Program 7.1 may produce odd results when $F(x)$ is close to 0 or 1. For instance, our microcomputer produced $F(5.597) = 0.99999999$ and $F(5.598) = 0.99999998$. The program is then improved by putting $F(x) = 0.99999888 + 0.0000002\,x$ for $5.3 < x \leqslant 5.6$ and $F(x) = 0.00000112 + 0.0000002\,x$ for $-5.6 \leqslant x < -5.3$. Make similar tests and modifications to suit your computer. You may not need so many decimal places; find the largest EP and the smallest C1 such that $F(x)$ is always correct to four decimal places.

(7.2) Trying to be clever, we deleted lines 500–510 in Program 7.1 and inserted line 535: IF Z > C1 THEN Y = 0.5 + 0.5*SGN(K/ABS(K)): GOTO 570. This failed for large z; how and why?

(7.3) Carry out tests of the normal approximation to the binomial distribution, particularly for values of p close to 0 or 1. For this purpose, you may prefer to add 10 000 to all the line numbers of Program 7.1 and then attach its 'menu' to Program 6.1 so that all the programs of Chapters 6 and 7 are run together. Alternatively, put Program 6.2 and parts of Program 7.1 into a special program which tabulates the binomial probabilities and the normal approximations in columns for comparison.

(7.4) Develop Program 7.1 so that it will tabulate the cumulative probabilities for a range of values, and the differences between these cumulative probabilities to give the probabilities for successive intervals, when the total range to be covered and the width of interval are inputted.

Chapter 8

Sampling and estimation

8.1 Introduction to statistical inference

We shall henceforth concentrate on statistical programs in BASIC and program notes, with simply a checklist of the statistical ideas which underlie them. An exception will be made in the case of regression and correlation, where fuller explanation of the underlying ideas is needed.

It is appropriate to repeat the warning given in the notes to the programs on discrete random variables: anyone who does not understand the explanations (including all the items on our 'checklist') can still operate the programs but is clearly incompetent to interpret the results.

Statistical inference consists of *estimation* and *hypothesis testing,* but it is first necessary to say something about sampling distributions.

8.2 Sampling distributions

We have hitherto treated the concepts 'sample mean' and 'sample standard deviation' as properties of sets of data. But since each of these statistics is defined by a specific formula which may just as readily be applied to random variables before an experiment is performed as to the collected data, we may regard them also as random variables. For instance, if $X_1, X_2, \ldots, X_n$ are independent identically-distributed random variables then the sample mean $\overline{X}$ is a random variable defined by:

$$\overline{X} = \frac{1}{n}(X_1 + X_2 + \ldots + X_n)$$

The probability distribution of a statistic is termed a *sampling distribution,* and is usually deduced from the probability distributions of the individual random variables on which the statistic is defined.

The term *expectation* (or 'expected value') is virtually a synonym for mean, but always as the mean of a theoretical distribution and not of a set of data. The *expectation of the mean,* $\mu_{\bar{X}}$, is the mean of the sampling distribution of the sample mean and is always equal to μ, the mean of the probability distribution of the individual random variables. The expectation of the sum of n identically-distributed random variables is $n\mu$. The expectation of $X+Y$ and of $X-Y$, where X and Y have means μ_X and μ_Y, are respectively $(\mu_X+\mu_Y)$ and $(\mu_X-\mu_Y)$.

The most useful formulae for the variances of statistics are applicable only when the random variables involved are independent. Then the variance of the mean is σ^2/n and the variance of the sum is $n\sigma^2$ (since multiplying the mean by n will multiply the variance by n^2). The variances of $X+Y$ and of $X-Y$ are both equal to $\sigma_X^2+\sigma_Y^2$, where σ_X^2 and σ_Y^2 are the individual variances.

The standard deviation of a statistic is termed a *standard error.* The *standard error of the mean* is therefore $\sigma/\sqrt{n}$, while the standard error of the sum of n independent identically-distributed random variables is $\sqrt{n}$ times σ. Since the binomial, negative binomial, and Poisson distributions can all be regarded as the distributions of sums of independent identically-distributed random variables, the formulae for their means and standard deviations can be seen to comply with these necessary properties.

If X and Y are independent, the *standard error of the sum* and the *standard error of the difference* are both equal to

$$\sqrt{\sigma_X^2+\sigma_Y^2}$$

In some cases we can say something more definite about the actual form of the sampling distribution as well as its parameters. For sums of independent identically-distributed random variables, we have seen that the Bernoulli, geometric, Poisson, negative exponential, normal, and chi-square variables lead respectively to sampling distributions which are binomial, negative binomial, Poisson, gamma, normal, and chi-square. Sums and differences of normally-distributed random variables will have normal distributions provided that the individual random variables are independent of each other, even if their means and variances differ.

Fundamental to the whole corpus of statistical inference, however, is the Central Limit Theorem, which indicates that the mean of a sufficiently-large number of independent identically-distributed random variables from *any* distribution will have an approximate normal distribution with mean μ and standard error $\sigma/\sqrt{n}$, where n is the number of random variables. It only remains for the reader to recognize that, since sampling without replacement from a large batch gives approximately the same results as sampling with replacement, the apparently-abstract discussion about means of independent random variables is directly relevant to statistical inference employing the means of random samples.

8.3 Point estimators

An *estimator* is a statistic which is used to estimate a parameter of a probability distribution. The definition links the statistic with the parameter; for instance, it would be meaningless to say that the sample mean is 'unbiased' unless and until we stipulate what parameter it is being used to estimate.

The standard methods of arriving at estimators are (i) the method of moments, (ii) maximum likelihood estimation, and (iii) the method of least squares.

Having arrived at estimators, we may examine their properties. Desirable properties, applicable in greater or lesser degree according to the circumstances, are (i) unbiasedness, (ii) consistency, (iii) minimum mean absolute error, (iv) minimum mean square error, and (v) maximum probability of being exactly correct. The last of these, applicable only to discrete random variables, is overlooked in most textbooks.

The definitions of these methods and properties may be found in textbooks. Unfortunately many students learn the definitions but forget the context. Maximum likelihood is not a 'desirable property' of estimators; neither is 'minimum variance'. It can however be shown that the minimum mean square error estimator among unbiased estimators, though not the minimum mean square error estimator among all estimators, will be the one with minimum variance.

In fact, this compromise of minimizing mean square error subject to maintaining unbiasedness is almost invariably adopted. It leads almost always to the simple rule that we use the sample mean $\overline{X}$ to estimate the mean μ of a probability distribution, and we use the sample standard deviation S to estimate the standard deviation σ of a probability distribution, where:

$$S^2 = \frac{1}{n-1}\sum_{i=1}^{n}(X_i-\overline{X})^2.$$

S^2 is then an unbiased estimator of σ^2, but S is not an unbiased estimator of σ.

8.4 Confidence interval for a population mean

If we were told that the estimated mean diameter of bolts in a batch was 15.4 mm, we might still be keen to know how much confidence to place in this figure. Were we to learn that the estimate was based on measuring the diameter of a single randomly-selected bolt, we would not feel very confident about rejecting the batch if the specification was for a mean diameter of 15 mm with a standard deviation not exceeding 0.3 mm. On the other hand, we would be fairly sure that the batch was unsatisfactory if 15.4 was the mean diameter in mm of a random sample of 100 bolts.

We put a figure on our uncertainty by using the observed data to produce a *confidence interval* for the unknown parameter, which in this example would be the mean diameter of bolts in the batch. Calling this μ, and assuming that we know σ to be 0.3, it follows that (assuming the diameters to be normally distributed) there is a probability of 0.025 that a single observation will be less than $\mu - 1.96\sigma$ and a probability of 0.025 that it will be greater than $\mu + 1.96\sigma$. So if we decide to define the interval from $X - 1.96\sigma$ to $X + 1.96\sigma$ as the interval in which we believe μ to lie, there is a probability of 0.95 that the interval obtained will in fact contain μ. This is called a *95% confidence interval* for μ.

If the actual observed diameter is 15.4 mm, the 95% confidence interval for μ based on $\sigma = 0.3$ becomes the interval from 14.812 to 15.988 mm. It would *not* be correct to say that there is a 'probability of 0.95' that μ is between 14.812 and 15.988, since μ is a parameter and not a random variable; the probabilistic statements can be applied only to the random interval as expressed in terms of the random variable X, and not to any specific interval based on an observed value.

For a 95% confidence interval based on a sample of size n we use the standard error of the mean, $\sigma/\sqrt{n}$, in place of σ. Not surprisingly, we obtain a narrower confidence interval than from a single observation. If $n = 100$, $\bar{x} = 15.4$, $\sigma = 0.3$, then the standard error of the mean is $\sigma_{\bar{X}} = 0.03$ and so the 95% confidence interval for the population mean is from 15.3412 to 15.4588.

The figure 1.96 arose from the fact that $\Phi(-1.96) = 0.025$ and $\Phi(1.96) = 0.975$. For a 99% confidence interval we find the value x such that $\Phi(x) = 0.995$, and this proves to be 2.5758.

Program 8.1 finds both the 95% and the 99% confidence interval for the population mean using the population standard deviation σ and the sample mean $\bar{x}$. If only a single observation is used, one simply inputs the number of observations as 1. If there are several observations but the mean has not been calculated, the program will calculate the mean and standard deviation of the sample before calculating the confidence intervals.

Confidence intervals may be obtained even when the population standard deviation is unknown, by making use of the sample standard deviation. This introduces an additional source of uncertainty, but the uncertainty has been precisely quantified in the t-distribution. A sample of size 30 has 29 'degrees of freedom' in the sample standard deviation, and it is found that there is then a probability of 0.025 that the sample mean will be less than $\mu - 2.04\ s/\sqrt{n}$ and a probability of 0.025 that it will be greater than $\mu + 2.04\ s/\sqrt{n}$. So we substitute s for σ and 2.04 for 1.96 to obtain the 95% confidence interval for μ.

For samples of size greater than 30, the correct figure is less than 2.04 and tends to 1.96 as the sample size tends to infinity. The program remains on the safe side by using 2.04 for $n \geqslant 30$ and drawing attention to the fact that it is an *approximate* confidence interval. It is the fact that the *t*-distribution is close to the normal distribution when $n \geqslant 30$ which has presumably led to the mistaken use of $n \geqslant 30$ for the normal approximation for the sample mean 'regardless of the shape of the population' as discussed in Section 7.3.

For samples of size 6 the correct figure for a 95% confidence interval using the sample standard deviation is about 2.58, while for smaller samples it increases to become 12.71 for samples of size 2. The program uses 2.04 and 2.58 to provide alternative versions when $6 \leqslant n \leqslant 29$ but does not attempt to produce an explicit confidence interval if $n < 6$.

Program 8.1.

```
100 REM CONFIDENCE INTERVAL WITH KNOWN S.D. OR LARGE SAMPLE
110 PRINT: PRINT: PRINT " CONFIDENCE INTERVAL FOR THE MEAN"
120 PRINT: PRINT "(Assuming an approximate Normal distn.)"
130 PRINT: PRINT "Input number of observations ";
    : READ N: PRINT ;N
140 PRINT: PRINT "Is the population standard deviation
    known?  (Y/N) ";
150 READ K$: PRINT K$
160 IF K$="Y" THEN PRINT "Input the population s.d. ";
    : READ SD: PRINT ;SD
170 PRINT: PRINT "Has the sample mean already been
    calculated?  (Y/N) ";
180 READ M$: PRINT M$
190 IF M$="Y" THEN PRINT "Input sample mean ";
    : READ M: PRINT ;M
200 IF K$="Y" AND M$="Y" GOTO 450
210 PRINT: PRINT "Has the sample standard deviation"
220 PRINT "already been calculated?  (Y/N) ";
    : READ S$: PRINT S$
230 IF S$="Y" THEN PRINT "Input sample s.d. ";
    : READ S: PRINT ;S
240 IF M$="Y" AND S$="Y" GOTO 320
250 PRINT: S1 = 0: S2 = 0
260 FOR I = 1 TO N
270 PRINT "Input observation No. ";I;": ";
    : READ X: PRINT ;X
280 S1 = S1+X: S2 = S2+X*X
```

```
290 NEXT I
300 M = S1/N: PRINT: PRINT "Sample Mean = ";M
310 S = SQR((S2-S1*M)/(N-1)): PRINT "Sample Standard
    Deviation = ";S
320 SE = S/SQR(N): PRINT: PRINT "Estimated Standard Error
    of the Mean = ";SE
330 IF K$="Y" GOTO 450
340 IF N<6 GOTO 410
350 PRINT: PRINT "APPROXIMATE CONFIDENCE ";: C = 2.04
360 IF N>29 THEN PRINT "INTERVALS:": D = 2.75: GOTO 470
370 PRINT "INTERVAL:": PRINT
380 PRINT "95% confidence interval for the popula-"
390 PRINT "tion mean is wider than ";M-C*SE;" to ";M+C*SE
400 L = M-2.58*SE: PRINT "and narrower than ";L;" to "; 2*M-L
410 PRINT: PRINT "For accurate confidence intervals a"
420 PRINT "table of the t-distribution should be"
430 PRINT "consulted.  Degrees of freedom = ";N-1
440 GOTO 110
450 SE = SD/SQR(N): PRINT: PRINT "Standard Error of the Mean = ";SE
460 C = 1.96: D = 2.5758
470 PRINT: PRINT "95% confidence interval for population
    mean = ";
480 L = M-C*SE: PRINT ;L;" to ";2*M-L
490 PRINT: PRINT "99% confidence interval for population
    mean = ";
500 L = M-D*SE: PRINT ;L;" to ";2*M-L
510 GOTO 110
600 DATA 1,Y,0.3,Y,15.4
610 DATA 100,Y,0.3,Y,15.4
620 DATA 9,Y,40,Y,1323
630 DATA 36,N,Y,22.5,Y,1.5
640 DATA 12,N,Y,42,Y,11.9
650 DATA 16,N,N,N
651 DATA 1.502,1.501,1.504,1.498,1.501,1.502,1.500,1.501
652 DATA 1.501,1.502,1.503,1.501,1.499,1.500,1.505,1.501
999 END
```

Program notes

(1) This program, and all subsequent programs, have been designed to READ . . . DATA and print the results out. As originally developed on a microcomputer it required additional lines to halt the program so that the screen could be read:

```
325 PRINT : PRINT "Press C to continue."
326 GET A$ : IF A$ = "" GOTO 326
```

or 326 A$ = GET$, with similar lines at 435–6 and 505–6. Further additional lines would be needed if it were desired to read the observations being inputted and there were more than 15 observations. (These requirements are for a screen with 25 rows × 40 columns.)

(2) The progam should be fairly self-explanatory. Note that if the answers to the first two questions, at lines 140 and 170, are both Y then the third question (at lines 210–220) is not asked. If the answers are N, Y, Y, the actual observations are not required. If they are N, N, Y, the sample mean has to be calculated and the program in fact computes the standard deviation also and ignores the value which

has been inputted. If they are Y, N, N, or Y, N, Y, the calculations proceed up to and including the 'Estimated Standard Error of the Mean' but then the program uses the true population standard deviation to calculate the confidence intervals.

(3) Note that the DATA always begins with the value of n. There follow the answers to two questions, a Y answer being followed in each case by the input of the parameter or statistic concerned. If the answers are both Y no more data are needed; otherwise the third question is asked and again a Y must be followed by a number. If the last two answers were not both Y the n observations must follow.

8.5 Other confidence intervals

If the population mean μ is known but the population standard deviation σ is unknown, a confidence interval for σ may be obtained using the sum of squared deviations from μ and the chi-square distribution with n degrees of freedom. If, as is more likely, the values of μ and σ are both unknown, a confidence interval for σ may be obtained using the sample standard deviation and the chi-square distribution with $(n-1)$ degrees of freedom.

In both these cases, we must be able to assume that the individual observations come from a normal distribution. (The fact that the sample mean may have an approximate normal distribution because n is large is irrelevant.)

It is also possible to produce a confidence interval for a proportion; in other words, a confidence interval for the unknown parameter p in a Bernoulli distribution. This may be calculated precisely or may be an approximate confidence interval based on the approximate normal distribution applicable where the parameters satisfy the requirements stipulated in Section 7.4.

We have only discussed confidence intervals which are symmetrical in probability. It would be possible if required to produce a 95% confidence interval starting at $-\infty$ for a population mean or at zero for a standard deviation or a proportion, so that the 5% of confidence intervals which 'miss' the true parameter all understate it. Again, we could produce 95% confidence intervals which terminate at the maximum possible value for the parameter and so can 'miss' the parameter only by overstating it.

Finally, we sometimes specify a 99.9% confidence interval to be sure 'beyond all reasonable doubt' that we will include the true value of the parameter. For the symmetrical confidence interval for the population mean this requires the value 3.2905 in place of 1.9600 or 2.5758.

There is no room in this text to do more than list these possibilities.

PROBLEMS

(8.1) Use Program 8.1 to calculate the confidence intervals for the population means for all examples you can find in your statistics textbook.

(8.2) Inventing sets of data is much harder work than getting the computer to invent them for you! The following program will generate almost-random observations to suit a specified target mean and standard deviation and with an approximate normal distribution.

```
10000 DIM D(255), N(4)
10010 NS = 100: M=20:S=4
10020 INPUT "Input number of observations to be simulated ";N
10030 INPUT "Input target mean, standard deviation ";MN,SD
10040 A=SD/S: B=MN-A*M:N1=N-1: NI=0
10050 FOR I=0 TO INT (N1/5)
10060 FOR K=1 TO NS: R=INT(5*RND(1)): N(R)=N(R)+1:
      NEXT K
10070 FOR J=0 TO 4: D(NI+J) = A*N(J)+B: N(J)=0:
      NEXT J
10080 NI=NI+5
10090 NEXT I
10100 S1=0: S2=0
10110 FOR I=0 TO N1: X=D(I): PRINT X;: S1=S1+X:
      S2=S2+X*X: NEXT I
10120 PRINT: PRINT MN;S1/N,SD;SQR ((S2-S1*S1/N)/N1)
10130 GOTO 10020
```

This simulates 100 throws of a five-sided die in rather the same way that Program 4.2 simulated a six-sided die. Each possible value from 0 to 4 has a $b(100,0.2)$ distribution, which has mean 20 and standard deviation 4. Five observations are produced at a time, but they are not truly random since they always total 100; this is unlikely to matter much for the present purpose, and has the advantage that if N is a multiple of 5 the simulated observations will have mean equal to exactly the target mean. The target standard deviation should preferably be 4, so that the observed values are all integers, or 4 times a power of 10 or of 0.1.

The observations can be produced more quickly by putting $NS=25$, $M=5$, $S=2$, but the fit to the normal distribution is much poorer and the fact that no observation can be more than 2.5 standard deviations below the mean is not entirely satisfactory. If these values are used for the three constants NS, M, and S, it is preferable to make the target standard deviation 2, or 2 times a power of 10 or of 0.1.

This program may be added as a subroutine to any statistics program which requires data which have an approximately normal distribution, being terminated at 10100 RETURN. Modify Program 8.1 to incorporate this subroutine and to use data from the array *D* as input at line 270. Further modify it so that it makes use of only one observation from each simulated set of five, and thereby simulates genuinely independent observations.

(8.3) Using random observations from a population with known mean and standard deviation, or independent observations simulated as in Problem 8.2, calculate the confidence intervals for a large number of samples and see what proportion of these intervals include the population mean.

Chapter 9

Hypothesis testing

9.1 Introduction to hypothesis testing

In Section 8.4 we discussed how a confidence interval for the mean diameter of the bolts in a batch might lead to a decision whether or not to reject the batch. If we knew that the specification was for a mean diameter of 15 mm and a standard deviation of 0.3 mm, the problem could have become one of testing a hypothesis.

If we were able to assume that the diameters were normally distributed around an unknown mean μ with standard deviation 0.3, we would formulate the *null hypothesis* that $\mu = 15$. There would then be a probability of 0.95 that a single observation would be between 14.412 and 15.588, if the null hypothesis were true. The test would proceed by selecting a specimen at random and measuring its diameter; if the diameter was between 14.412 and 15.588 we would accept the null hypothesis, and if not we would reject it. This would be a test with a 5% *significance level,* and would be described as a *two-tail* test.

Similarly, if the test were to be based on a sample of size 100, the standard error of the mean would be 0.03 and so we would accept the null hypothesis if the sample mean proved to be between 14.9412 and 15.0588 and reject it otherwise. It will be noted that the width of the 'acceptance interval' for the 5% two-tail test is in each case the same as the width of the 95% confidence interval for the population mean. Hence the test leads to acceptance of the null hypothesis if the observed value is such that the confidence interval for μ includes 15 (as for a value of 15.4 for a single observation), and to rejection of the null hypothesis if the confidence interval does not include 15 (as for a sample mean of 15.4 for a sample of size 100). This simple relationship between a 95% confidence interval which is symmetrical in probability and a 5% two-tail test is equally applicable between a 99% confidence interval and a 1% two-tail test or between a 99.9% confidence interval and a 0.1% two-tail test.

However, not all hypothesis tests can be seen simply as the converse of determining confidence intervals. It would be difficult to find a 'confidence interval' problem which is the converse of the goodness-of-fit test we shall briefly consider in Chapter 11. The principles of hypothesis testing are best understood by learning the five or six basic definitions, which are so interrelated that it is necessary to study them all simultaneously.

A *statistical hypothesis* is 'a statement which implies that the true probability distribution describing the inherent variability in an observational situation belongs to a proper subset of the family of possible probability distributions' (S. D. Silvey). The hypothesis which is 'under test' is called the *null hypothesis*. The negation of the null hypothesis is the *alternative hypothesis*.

A *test of a statistical hypothesis* is a procedure for deciding whether to accept or reject the hypothesis. The *critical region* (or 'rejection region') of the test is that part of the sample space which leads to rejection of the null hypothesis. The *significance level* of the test is the probability that the test will lead to rejection of the null hypothesis when the null hypothesis is true; that is, it is the probability under the null hypothesis that the result will fall into the critical region.

The act of rejecting the null hypothesis when it is true is called a *type I error*. So the significance level is the probability of a type I error. The act of accepting the null hypothesis when it is false is called a *type II error*.

If the result of the test is in the critical region, we reject the null hypothesis and accept the alternative hypothesis. So the critical region should be located where it will suit the alternative hypothesis. The null hypothesis should be, in comparison with the alternative hypothesis, the hypothesis which it is the more serious wrongly to reject.

9.2 Test of a population mean

Program 9.1 accepts data in a similar manner to Program 8.1, except that it also enquires for the 'population mean under the Null Hypothesis'. Since the program also produces confidence intervals, it will accept the value 1E + 11 as an indication that there is no Null Hypothesis and only the confidence intervals are required.

If the population standard deviation is known, the program calculates the standard normal variable $z = (x - \mu)/\sigma$ as discussed in Section 7.3; but note that if the sample size exceeds 1 the calculation uses $\bar{x}$ in place of x and $\sigma/\sqrt{n}$ in place of σ, and that the value of μ employed is that stipulated by the null hypothesis. Thus if $\mu = 15$ under the null hypothesis, $n = 100$, $\sigma = 0.3$, and $\bar{x} = 15.4$, z will work out at $13\frac{1}{3}$; if $n = 1$ and $x = 15.4$, z will work out at $1\frac{1}{3}$. The program then uses the algorithm for computing $\Phi(z)$ to produce the 'upper-tail probability', which is the probability under the null hypothesis of obtaining a result greater than that actually obtained. The upper-tail probabilities will therefore be respectively $1-\Phi\ (13\frac{1}{3})$ which is effectively zero, and $1-\Phi\ (1\frac{1}{3})$ which is 0.09121.

This means that in the former case the result would be 'significant' at any level, though 0.1% is the most stringent level commonly used. In the latter case the result would be 'significant' and hence lead to rejection of the null hypothesis only if the significance level was greater than 0.09121 in an 'upper-tail test'; the least stringent level commonly used is 5% and so we would not reject the null hypothesis. The two-tail probability is also printed out, and would be applied if the test were a two-tail test. It is necessary for the user to know the form of the null and the alternative hypotheses and hence to know whether a one-tail or a two-tail test is appropriate. For a one-tail test he must also know *which* tail is appropriate for the alternative hypothesis; if z is negative the program prints out the lower-tail probability, but this cannot lead to rejection of the null hypothesis however small the probability proves to be if the test is an upper-tail test.

If the population standard deviation is unknown, Program 9.1 uses the sample standard deviation to calculate the 'Estimated Standard Error of the Mean'. The t-distribution is then appropriate, as already discussed in Section 8.4. However, if $n \geqslant 30$ the t-distribution is sufficiently close to the normal distribution for the latter to be used; this is known as a 'large sample test'. If $n < 30$ the program calculates the t statistic, which has the same formula as the z statistic except that s is used in place of σ, prints out the number of degrees of freedom which is $(n-1)$, and leaves it to the user to complete the test by looking up a table of the t-distribution. The term 'small sample test' is used for a test involving a sample of less than 30 observations in which the population standard deviation is unknown and the t-distribution therefore has to be referred to.

Program 9.1

```
100 EP = 1E-9: C1 = 5.6: C2 = 1/SQR(2*PI): C3 = 1.96
    : C4 = 2.5758
110 PRINT: PRINT: PRINT "HYPOTHESIS TEST AND CONFIDENCE
    INTERVAL"
120 PRINT: PRINT "(Assuming an approximate Normal distn.)"
130 PRINT: PRINT "Input number of observations ";
    : READ N: PRINT ;N
140 PRINT: PRINT "Input population mean under the Null
    Hypothesis.  If only ";
150 PRINT "a confidence interval is required,
    input 1E+11. ";
160 READ MO: PRINT ;MO
170 PRINT: PRINT "Is the population standard deviation
    known?  (Y/N) ";
180 READ K$: PRINT K$
190 IF K$="Y" THEN PRINT "Input population s.d. ";
    : READ SD: PRINT ;SD
200 PRINT: PRINT "Has the sample mean already been
    calculated?  (Y/N) ";
```

```
210 READ M$: PRINT M$
220 IF M$="Y" THEN PRINT "Input sample mean ";
    : READ M: PRINT ;M
230 IF K$="Y" AND M$="Y" GOTO 410
240 PRINT: PRINT "Has the sample standard deviation"
250 PRINT "already been calculated?  (Y/N) ";
    : READ S$: PRINT S$
260 IF S$="Y" THEN PRINT "Input sample s.d. ";
    : READ S: PRINT ;S
270 IF M$="Y" AND S$="Y" GOTO 350
280 PRINT: S1 = 0: S2 = 0
290 FOR I = 1 TO N
300 PRINT "Input observation No. ";I;": ";
    : READ X: PRINT ;X
310 S1 = S1+X: S2 = S2+X*X
320 NEXT I
330 M = S1/N: PRINT: PRINT "Sample Mean = ";M
340 S = SQR((S2-S1*M)/(N-1)): PRINT "Sample Standard
    Deviation = ";S
350 SE = S/SQR(N): PRINT: PRINT "Estimated Standard Error
    of the Mean = ";SE
360 IF K$="Y" GOTO 410
370 IF N>29 GOTO 420
380 IF MO<1E+10 THEN PRINT: PRINT "Small Sample Test:
    t = ";(M-MO)/SE
390 PRINT: PRINT "Degrees of freedom = ";N-1
400 GOTO 110
410 SE = SD/SQR(N): PRINT: PRINT "Standard Error of the
    Mean = ";SE
420 IF MO>1E+10 GOTO 550
430 Z = (M-MO)/SE: PRINT: PRINT "Null Hypothesis z value
    = ";Z
440 IF Z<-C1 THEN Y = 0: GOTO 510
450 IF Z>C1 THEN Y = 1: GOTO 510
460 ZZ = Z*Z: K = C2*EXP(-ZZ/2)
470 IF Z<0 THEN Z = -Z: K = -K
480 Y = Z: I = 1
490 I = I+2: Z = Z*ZZ/I: Y = Y+Z: IF Z>EP GOTO 490
500 Y = 0.5+K*Y
510 PRINT: IF K<=0 THEN PRINT "Lower-";
520 IF K>0 THEN PRINT "Upper-";: Y = 1-Y
530 PRINT "tail probability = ";Y
540 PRINT "Two-tail probability = ";2*Y
550 PRINT: PRINT "95% confidence interval for population
    mean = ";
560 L = M-C3*SE: PRINT ;L;" to ";2*M-L
570 PRINT: PRINT "99% confidence interval for population
    mean = ";
580 L = M-C4*SE: PRINT ;L;" to ";2*M-L
590 GOTO 110
600 DATA 1,15,Y,0.3,Y,15.4
610 DATA 100,15,Y,0.3,Y,15.4
620 DATA 9,1300,Y,40,Y,1323
630 DATA 36,1E+11,Y,1.5,Y,22.5
640 DATA 12,50,N,Y,42,Y,11.9
650 DATA 16,1.5,N,N,N
651 DATA 1.502,1.501,1.504,1.498,1.501,1.502,1.500,1.501
652 DATA 1.501,1.502,1.503,1.501,1.499,1.500,1.505,1.501
999 END
```

Program notes

(1) For discussion of EP, C1 and C2 at line 100, see note (1) to Program 7.1 in Section 7.3.

(2) If the output has to be read from a screen, additional lines will be needed as discussed in note (1) to Program 8.1. The recommended

places for these additional lines are 275–276, 365–366, 395–396, 415–416, and 585–586.

(3) The manner in which the program asks for and accepts data is similar to that discussed in notes (2) and (3) to Program 8.1, except that the 'population mean under the Null Hypothesis' (or 1E+11) has to be inputted immediately after the sample size n.

(4) If n is 1 and the population standard deviation is unknown, no test is possible. The program will produce a DATA ERROR at line 340 when it tries to divide by zero (see Problems).

(5) Lines 440–500 apply the same algorithm to find $\Phi(z)$ as lines 500–560 of Program 7.1.

9.3 Test for equality of population means

A common problem in statistical inference is to test the hypothesis that two populations have the same mean. We may wish to decide whether or not a new method of producing yarn or wire leads to an increase in the mean breaking load, or whether a new medical treatment leads to a decrease in the time to recovery. We carry out tests on breaking loads or on times to recovery, using the 'old' method for one sample and the 'new' method for a second sample, and test the hypothesis that there is no difference between the means of the populations from which the two samples are drawn.

The test statistic is the difference between the two sample means, and so the relevant sampling distribution is that of $\overline{Y}-\overline{X}$. The expectation of this statistic is $\mu_{\overline{Y}}-\mu_{\overline{X}}$ which is $\mu_Y-\mu_X$ and so is zero under the null hypothesis. The standard error of the test statistic is obtained by:

$$\text{S.E.} = \sqrt{\sigma_{\overline{X}}^2+\sigma_{\overline{Y}}^2} = \sqrt{\frac{\sigma_X^2}{n_X}+\frac{\sigma_Y^2}{n_Y}}$$

using first the result for the standard error of the difference (since $\overline{X}$ and $\overline{Y}$ can be assumed to be independent) and then the result for the standard error of the mean for a sample of n_X observations on X and for a sample of n_Y observations on Y.

This standard error becomes the denominator of the z statistic, and the numerator is simply $(\overline{y}-\overline{x})$, the observed difference between the sample means; the term $-\mu$ disappears since the expectation of the difference between the sample means is zero under the null hypothesis. The one-tail and two-tail probabilities are then computed as for the test of a population mean in Section 9.2, and the decision can be made as to whether to accept or reject the null hypothesis. In the cases considered above we would have an upper-tail test for the difference in breaking loads and a lower-tail test for the difference in times to recovery, since in each case we would only be interested in the alternative hypothesis that the 'new' method produces an improvement, and the possibility that it may actually be worse than the 'old' method would be amalgamated into the null hypothesis of 'no improvement'.

If the population standard deviations are unknown, the test can proceed using the sample standard deviations instead provided that *both* samples are of size 30 or over.

If the samples are 'small' and σ is unknown for one or both samples, there can be a fairly straightforward test using the t-distribution only if the variances are 'compatible'; that is, only if we can reasonably conclude that the populations have the same variance.

This can be tested using the F-ratio test, in which the ratio of the two sample variances is subjected to test using the tables of the F-distribution with the appropriate numbers of degrees of freedom. This is a two-tailed test, but in practice the tables give only the upper-tail values; so we divide the larger sample variance by the smaller and use the table of 2½% upper-tail values for a 5% two-tail test.

For convenience, the F-ratio is calculated in Program 9.2 even in the case of two large samples where it is irrelevant to the test for equality of population means. The program proceeds with the calculation of the t statistic for a small sample test even if the F-ratio is significantly large so that the assumption underlying the t-test has to be rejected.

If the variances are compatible, the t-test requires that the denominator of the t statistic must be:

$$\text{S.E.} = \sqrt{\frac{s^2}{n_X} + \frac{s^2}{n_Y}} = s\sqrt{\frac{1}{n_X} + \frac{1}{n_Y}}$$

where s^2 is the estimate of the common variance of the two populations and is obtained by weighting the sample variances in proportion to their respective degrees of freedom:

$$s^2 = \frac{(n_X - 1)s_X^2 + (n_Y - 1)s_Y^2}{n_X + n_Y - 2}$$

The number of degrees of freedom for the test is $n_X + n_Y - 2$.

Program 9.2

```
100 EP = 1E-9: C1 = 5.6: C2 = 1/SQR(2*PI)
110 PRINT: PRINT: PRINT "TEST FOR EQUALITY OF POPULATION
    MEANS"
120 PRINT: PRINT "(Assuming an approximate Normal distn.)"
130 PRINT: PRINT "Input numbers of observations ";
    : READ NX,NY: PRINT ;NX;" ";NY
```

```
140 PRINT: PRINT "Are the population standard deviations
    known?  (Y/N) ";
150 READ K$: PRINT K$
160 IF K$="Y" THEN PRINT "Input popn. s.ds. ";
    : READ SX,SY: PRINT ;SX;" ";SY
170 IF K$="Y" THEN SE = SQR(SX*SX/NX+SY*SY/NY)
180 PRINT: PRINT "Have the sample means already been
    calculated?  (Y/N) ";
190 READ M$: PRINT M$
200 IF M$="Y" THEN PRINT "Input sample means ";
    : READ MX,MY: PRINT ;MX;" ";MY
210 IF K$="Y" AND M$="Y" GOTO 420
220 PRINT: PRINT "Have the sample standard deviations"
230 PRINT "already been calculated?  (Y/N) ";
    : READ S$: PRINT S$
240 IF S$="Y" THEN PRINT "Input sample s.ds. ";
    : READ SX,SY: PRINT ;SX;" ";SY
250 IF M$="Y" AND S$="Y" GOTO 310
260 PRINT: PRINT "First sample: ": N = NX: GOSUB 600
270 MX = M: SX = S
280 PRINT: PRINT "Second sample: ": N = NY: GOSUB 600
290 MY = M: SY = S
300 PRINT: IF K$="Y" GOTO 420
310 PRINT: PRINT "F-ratio = ";: VX = SX*SX: VY = SY*SY
    : F = VX/VY
320 IF F>=1 THEN PRINT ;F;" with ";NX-1;" ";NY-1;
    " degrees of freedom."
330 IF F<1 THEN PRINT ;1/F;" with ";NY-1;" ";NX-1;
    " degrees of freedom."
340 IF NX>29 AND NY>29 GOTO 410
350 S = SQR(((NX-1)*VX+(NY-1)*VY)/(NX+NY-2))
360 PRINT: PRINT "Pooled Standard Deviation = ";S
    : SE = S*SQR(1/NX+1/NY)
370 PRINT: PRINT "Estimated Standard Error of Difference
    of Means = ";SE
380 PRINT: PRINT "Small Sample Test:  t = ";(MY-MX)/SE
390 PRINT: PRINT "Degrees of freedom = ";NX+NY-2
400 GOTO 110
410 SE = SQR(VX/NX+VY/NY): PRINT "Estimated ";
420 PRINT "Standard Error of Difference of Means = ";SE
430 Z = (MY-MX)/SE: PRINT: PRINT "Null Hypothesis z value
    = ";Z
440 IF Z<-C1 THEN Y = 0: GOTO 510
450 IF Z>C1 THEN Y = 1: GOTO 510
460 ZZ = Z*Z: K = C2*EXP(-ZZ/2)
470 IF Z<0 THEN Z = -Z: K = -K
480 Y = Z: I = 1
490 I = I+2: Z = Z*ZZ/I: Y = Y+Z: IF Z>EP GOTO 490
500 Y = 0.5+K*Y
510 PRINT: IF K<=0 THEN PRINT "Lower-";
520 IF K>0 THEN PRINT "Upper-";: Y = 1-Y
530 PRINT "tail probability = ";Y
540 PRINT "Two-tail probability = ";2*Y
550 GOTO 110
600 S1 = 0: S2 = 0
610 FOR I = 1 TO N
620 PRINT "Input observation No. ";I;": ";
    : READ X: PRINT ;X
630 S1 = S1+X: S2 = S2+X*X
640 NEXT I
650 M = S1/N: PRINT: PRINT "Sample Mean = ";M
660 S = SQR((S2-S1*M)/(N-1)): PRINT "Sample Standard
    Deviation = ";S
670 RETURN
700 DATA 36,36,Y,3,4,Y,16,17.25
710 DATA 12,10,Y,4,5,Y,85,81
720 DATA 12,10,N,Y,85,81,Y,4,5
730 DATA 12,12,Y,11,11,N,N
```

```
731 DATA 39,43,43,52,52,59,40,45,47,62,40,27
732 DATA 42,37,61,74,55,57,44,55,37,70,52,55
740 DATA 5,5,N,N,N
741 DATA 10,16,19,27,32,34,21,12,30,19
750 DATA 7,5,N,N,N
751 DATA 12,15,11,16,14,14,16
752 DATA 8,10,14,10,13
999 END
```

Program notes

The program can readily be understood by comparing it with Program 9.1 in the light of the explanations given in the text. If the output has to be read from a screen, the recommended places for the lines to halt the display are 245–246, 275–276, 305–306, 395–396, 425–426, and 545–546.

9.4 Test of matched samples

Sometimes an experiment can be planned so that two 'matched samples' are involved. For instance, 12 specimens of coated steel pipe are buried in different soils, each with a specimen of uncoated steel pipe alongside it, to compare the corrosion. The results in matched pairs are:

Coated	39	43	43	52	52	59	40	45	47	62	40	27
Uncoated	42	37	61	74	55	57	44	55	37	70	52	55

Test whether the results support the claim that coated pipe is less liable to corrode.

This type of experiment is sometimes called the 'method of controls' or the 'test of paired comparisons'. It is essential that the data are kept in the correct pairs; thus the observation 39 for coated pipe is for the same soil as the figure 42 for uncoated pipe.

The test is carried out as if the results for one of the samples constitute a fixed measurement of the properties of the items which are subjected to the experiment (in this case, the properties of the 12 different soils) and it is in each case the *difference* between this fixed measurement and the measurement using the alternative treatment which is the random variable being studied. So we interpret the data as providing 12 observed differences: 3, −6, 18, 22, 3, −2, 4, 10, −10, 8, 12, 28. The null hypothesis is that these are a random sample from a population with mean zero.

In this example, the mean difference is 7.5 and the standard deviation of the differences is 11.28. If the population standard deviation is unknown, as will almost always be the case, the t-test is used. The t statistic is $7.5/3.256 = 2.303$, with 11 degrees of freedom. An upper-tail test is required, and the result is significant at the 5%

level though not at the 1% level. We would conclude that the results support the claim.

The principles governing the 'matched samples' test may be re-emphasized by an artificial example:

Coated	10	16	19	27	32
Uncoated	12	19	21	30	34

These matched samples produce a t statistic of 9.798 with 4 degrees of freedom, which is significant at the 0.1% level, and establish the real effect of using coated pipe beyond all reasonable doubt. This is because there is a consistent improvement of 2 or 3 in every matched pair, whatever the nature of the soil.

Now consider two unmatched samples:

Coated	10	16	19	27	32
Uncoated	34	21	12	30	19

Tested correctly using Program 9.2, this produces a t statistic of 0.432 with 8 degrees of freedom, which is nowhere near being significant even at the 10% level.

If we had, quite incorrectly, treated the matched-sample data as if they were for unmatched samples, we would have got the same result as for the samples which were genuinely unmatched. This would be because the large variation among the soils would have produced a large estimate for s and so would have concealed the importance of the difference between the sample means (which remains at 2.4).

An even more serious error, perhaps better described as 'piece of deceit', would be to re-arrange the observations for the uncoated pipe in the unmatched samples into ascending order and then treat the samples as matched samples. Since the data would then be the same as for the samples which were genuinely matched, a completely false claim to have obtained a significant result would ensue.

Program 9.3

```
100 PRINT: PRINT: PRINT "  MATCHED SAMPLES TEST"
110 PRINT: PRINT "(Assuming an approximate Normal distn.)"
120 PRINT: PRINT "Input number of matched pairs ";
    : READ N: PRINT ;N
130 PRINT: S1 = 0: S2 = 0
140 FOR I = 1 TO N
150 PRINT "Input pair No. ";I;": ";: READ X,Y
    : PRINT ;X;" ";Y
160 D = Y-X: S1 = S1+D: S2 = S2+D*D
170 NEXT I
180 M = S1/N: PRINT: PRINT "Mean Difference = ";M
190 S = SQR((S2-S1*M)/(N-1)): PRINT "Standard Deviation of
    Differences = ";S
200 PRINT: PRINT "Test Statistic t = ";M*SQR(N)/S
210 PRINT: PRINT "Degrees of freedom = ";N-1
220 GOTO 100
```

```
300 DATA 12
301 DATA 39,42,43,37,43,61,52,74,52,55,59,57
302 DATA 40,44,45,55,47,37,62,70,40,52,27,55
310 DATA 5,10,12,16,19,19,21,27,30,32,34
320 DATA 8,35,36,37,41,30,33,27,26,33,35,31,36,29,36,35,35
999 END
```

Program notes

The program can readily be understood by comparing it with Program 9.1 in the light of the explanations given in the text. If the output has to be read from a screen, lines to halt the display are required as lines 215–216.

PROBLEMS

(9.1) Use Program 9.1 to carry out the following hypothesis tests:

(a) $n = 1$, null hypothesis $\mu = 15$, $\sigma = 0.3$, $x = 15.4$, two-tail test;

(b) $n = 1$, null hypothesis $\mu = 21$, $\sigma = 2$, $x = 16$, lower-tail test;

(c) $n = 100$, null hypothesis $\mu = 70$, $\sigma = 8.9$, $\bar{x} = 71.8$, upper-tail test;

(d) $n = 9$, null hypothesis $\mu = 1300$, $\sigma = 40$, $\bar{x} = 1323$, upper-tail test;

(e) $n = 100$, null hypothesis $\mu = 5$, σ unknown, $\bar{x} = 5.008$, $s = 0.036$, two-tail test;

(f) $n = 12$, null hypothesis $\mu = 50$, σ unknown, $\bar{x} = 42$, $s = 11.9$, two-tail test;

(g) $n = 10$, null hypothesis $\mu = 70$, σ unknown, observations are 73, 79, 71, 63, 85, 79, 61, 67, 83, 89, upper-tail test;

(h) $n = 16$, null hypothesis $\mu = 1.5$, σ unknown, observations are 1.502, 1.501, 1.504, 1.498, 1.501, 1.502, 1.500, 1.501, 1.501, 1.502, 1.503, 1.501, 1.499, 1.500, 1.505, 1.501, two-tail test.

(9.2) Modify Program 9.1 so that if $n = 1$ and σ is unknown it reverts to line 110 with an explanation instead of producing a DATA ERROR.

(9.3) Use Program 9.1 for all hypothesis tests and confidence intervals asked for in your statistics textbook for which it is appropriate.

(9.4) Use Program 9.2 to carry out the following tests for equality of population means:

(a) $n_X = 36$, $n_Y = 36$, $\sigma_X = 3$, $\sigma_Y = 4$, $\bar{x} = 16$, $\bar{y} = 17.25$, upper-tail test;

(b) $n_X = 16$, $n_Y = 14$, σ_X unknown, σ_Y unknown, $\bar{x} = 107$, $\bar{y} = 112$, $s_X = 10$, $s_Y = 8$, two-tail test;

(c) $n_X = 150$, $n_Y = 200$, σ_X unknown, σ_Y unknown, $\bar{x} = 1100$, $\bar{y} = 1120$, $s_X = 120$, $s_Y = 80$, two-tail test.

(9.5) Use Program 9.2 for all tests for equality of population means used as illustrations or asked for as problems in your statistics textbook.

(9.6) Eight specimens of acid are each divided into two. One half of each specimen is given to chemist A for testing its strength, while the other half is similarly tested by chemist B. Test whether the results, giving the numbers of cubic centimetres of standard alkali used, indicate that one chemist is prone to obtain higher results than the other:

A:	35	37	30	27	33	31	29	35
B:	36	41	33	26	35	36	36	35

(9.7) Twelve specimens of fabric are each divided into two, and one half of each is treated by a process which is claimed to increase the breaking load. The increases in breaking loads are recorded as:

$-1 \quad 7 \quad -2 \quad 9 \quad 11 \quad 4 \quad 7 \quad 1 \quad 10 \quad -2 \quad 3 \quad 11$

Is the change in mean breaking loads significant? (Hint: Use Program 9.1, null hypothesis $\mu = 0$, or use Program 9.3 inputting zero for the first value in each pair and the 'increase in breaking load' for the second value in each pair.)

(9.8) Use Program 9.3 for all tests of 'matched samples' or 'paired comparisons' used as illustrations or asked for as problems in your statistics textbook.

(9.9) Use the subroutine developed in Problem 8.2 to simulate sets of data for testing in Programs 9.1, 9.2, and 9.3.

Chapter 10

Regression and correlation

10.1 Straight line regression on a controlled variable

In this section we shall consider the problem of estimating the best straight line to relate a random variable to a controlled variable. For instance, in an agricultural experiment, the yield is found to depend on the amount of water supplied. The data, measured in suitable units, are

Water (x)	12	18	24	30	36	42	48
Yield (y)	5.27	5.68	6.25	7.21	8.02	8.71	8.42

Clearly there is a tendency for the yield to increase as the water increases, and the *straight line regression of Y on x* is the straight line which gives us the best estimate of how Y depends on x. It is determined using least squares estimation. This principle leads to the choice of the arithmetic mean as the estimator; however, the mean is now not a constant but a function of x:

$$\mu_Y(x) = \alpha + \beta x$$

Making use of the observed pairs of data, (x_i, y_i), we estimate α and β by finding the values of a and b such that the 'predicted' values y_i' given by:

$$y_i' = a + bx_i$$

are 'as close as possible' to the corresponding observed values y_i. On the principle of least squares, we choose a and b so as to minimize 'the sum of squared residuals',

$$\sum_{i=1}^{n} (y_i - y_i')^2$$

By expressing this sum as a function of a and b and equating the partial derivatives with respect to a and with respect to b to zero, it is found that the formulae are:

$$b = \frac{\sum_{i=1}^{n} x_i y_i - \bar{x} \sum_{i=1}^{n} y_i}{\sum_{i=1}^{n} x_i^2 - \bar{x} \sum_{i=1}^{n} x_i}$$

$$a = \bar{y} - b\bar{x}$$

There are other ways of arranging the formulae, but these are the simplest in practical computing and are applied in Program 10.1. For the above example they show that: $b = 0.1029$, $a = 3.994$. Hence the straight line regression of y on x is estimated as:

$$y' = 3.994 + 0.1029x$$

Thus when $x = 12$, the predicted $y' = 5.23$, and so the residual $(y-y')$ is 0.04. All the seven residuals may be calculated in this way, and the sum of squared residuals is 0.614. The 'total variation' among the values of y works out as 11.278, and hence the variance of the y values is 1.880. The 'explained variation' is $11.278-0.614=10.664$, and the ratio of this to the total variation is 0.946 which is known as the *coefficient of determination.* This coefficient takes the value 1 if the y values lie precisely on a straight line with positive or negative slope (but not zero slope), and is negligible if the values of y show no dependence on x.

It should be emphasized that in this type of problem the x values are controlled and not random, and so there is no such thing as 'correlation' between the x and y values.

The sources of error in estimating the regression line are (i) the assumption that the relationship is a straight line may not be correct; (ii) even if it *is* a straight line, the parameters may be inaccurately estimated if there are not many observations; (iii) even if there is a straight line correctly estimated, the random variability about the line may be so great that the line is not of much use; and (iv) there may be computational problems if the denominator in the formula for b is very small.

The first of these difficulties is best met by examining the residuals; if there is a run of negative residuals, then a run of positive residuals, and finally another run of negative residuals as x increases, this is an indication that the true relationship is a curve rather than a straight line. In any case, one would not trust the line to estimate y well below or well above the range of y values involved in estimating the line; we would certainly not conclude that supplying water (x) at the level 100 could be expected to produce a yield (y) of 14.28!

There are methods of testing the significance of the regression line (against the null hypothesis that there is no dependence of y on x) and of producing a confidence interval for the true parameters. The standard deviation of a predicted value, reflecting both the errors in estimating the line and the random variability about the line, can be calculated. We have not introduced these complexities into Program 10.1, but it produces warnings against unreliable estimates, and declines to produce any estimate if the denominator for b is less than a stipulated minimum value.

Program 10.1

```
100 REM STRAIGHT LINE REGRESSION ON A CONTROLLED VARIABLE.
110 EP = 1E-15: DIM X(255),Y(255)
120 PRINT: PRINT: PRINT "STRAIGHT LINE REGRESSION"
130 PRINT: PRINT "Input number of observations ";
    : READ N: PRINT ;N
140 IF N<2 GOTO 750
150 IF N=2 GOSUB 800
160 IF N>2 AND N<6 GOSUB 900
170 PRINT: PRINT "For each observation, input the value"
180 PRINT "of the explanatory (x) variable,"
190 PRINT "followed by the value of the dependent (y)
    variable.": PRINT
200 TX = 0: XX  = 0: TY = 0: YY = 0: XY = 0
210 XL = 1E+11: YL = XL: YH = -YL
220 FOR I = 1 TO N
230 PRINT "Input observation No. ";I;": ";
240 READ X: PRINT ;X;" ";: IF X<XL THEN XL = X
250 READ Y: PRINT ;Y: IF Y<YL THEN YL = Y
260 IF Y>YH THEN YH = Y
270 TX = TX+X: XX = XX+X*X: X(I) = X
280 TY = TY+Y: YY = YY+Y*Y: XY = XY+X*Y: Y(I) = Y
290 NEXT I
300 MX = TX/N: PRINT: PRINT "Mean of x = ";MX
310 X2 = XX-MX*TX: VX = X2/(N-1)
    : PRINT "Variance of x = ";VX
320 SX = SQR(VX): PRINT "Standard Deviation of x = ";SX
330 PRINT "Coefficient of Variation of x ";
    : IF XL>0 THEN PRINT "= ";SX/MX
340 IF XL<=0 THEN PRINT "does not exist."
350 MY = TY/N: PRINT: PRINT "Mean of y = ";MY
360 Y2 = YY-MY*TY: VY = Y2/(N-1)
    : PRINT "Variance of y = ";VY
370  SY = SQR(VY): PRINT "Standard Deviation of y = ";SY
380 PRINT "Coefficient of Variation of y ";
    : IF YL>0 THEN PRINT "= ";SY/MY
390 IF YL<=0 THEN PRINT "does not exist."
400 IF X2<EP THEN PRINT: PRINT "NO UNIQUE SOLUTION"
    : GOTO 120
410 Z = XY-MX*TY: B = Z/X2: A = MY-B*MX
420 PRINT: PRINT "The regression of Y on the explanatory"
430 PRINT "variable is estimated as:": PRINT
440 PRINT "Y = ";A;" ";: IF B>=0 THEN PRINT "+ ";
450 PRINT B;" x"
460 PRINT: PRINT "Coefficient of Determination ";
470 IF Y2<EP THEN PRINT "cannot be found.": GOTO 500
480 CD = Z*B/Y2: PRINT "= ";CD
500 PRINT: PRINT "Do you want to see the residuals?
    (Y/N) ": READ A$
510 IF A$="N" GOTO 580
520 RR = 0: FOR I = 1 TO N
530 X = X(I): Y = Y(I): Y1 = A+B*X: R = Y-Y1: RR = RR+R*R
540 PRINT ;X;TAB(6);Y;TAB(13);Y1;TAB(26);R
550 NEXT I
560 IF Y2<EP GOTO 580
570 PRINT: PRINT "Now C of D = ";1-RR/Y2
    : PRINT "Previously:  ";CD
580 PRINT: PRINT "Do you want to predict a new y value?
    (Y/N) ": READ A$
590 IF A$="N" GOTO 120
600 PRINT: PRINT "Input the value of the explanatory (x)"
610 PRINT "variable ";: READ X: PRINT ;X
620 Y = A+B*X: PRINT: PRINT "Predicted y = ";Y
630 IF Y<YL GOTO 700
640 IF Y>YH GOTO 720
650 GOTO 580
700 PRINT: PRINT "WARNING: This is below the range of the"
```

```
710 PRINT "observed y values and is unreliable.": GOTO 580
720 PRINT: PRINT "WARNING: This is above the range of the"
730 PRINT "observed y values and is unreliable.": GOTO 580
750 PRINT: PRINT "USELESS!  More observations needed."
760 GOTO 130
800 PRINT: PRINT "With only two observations the line"
810 PRINT "will fit the data exactly.  It will be"
820 PRINT "unreliable for predictive purposes.": RETURN
900 PRINT: PRINT "With so few observations, the estimated"
910 PRINT "regression equation will be unreliable.": RETURN
950 DATA 7,12,5.27,18,5.68,24,6.25,30,7.21,36,8.02
951 DATA 42,8.71,48,8.42,Y,Y,12,Y,24,Y,36,Y,48,N
999 END
```

Program notes

(1) Lines 140–160 provide for warnings if there are relatively few observations; see lines 750–910.

(2) Lines 170–290 collect the appropriate statistics from the data, including the minimum value of x and the minimum and maximum values of y.

(3) Lines 300–390 produce the statistics which may be of interest.

(4) Line 400 tests if the denominator for b is too small for reliable computation. It may be found that the value of EP, set at line 110, can be made smaller or should be made larger depending on the computer.

(5) Lines 410–480 calculate the regression line and the coefficient of determination, with another necessary check against EP for reliability.

(6) Lines 500–570 provide a printout of the residuals if desired. By calculating the coefficient of determination again from the residuals they provide a check on the numerical efficiency of the BASIC.

(7) Lines 580–650 invite new x values from which the predicted y can be calculated. Lines 630–640 and 700–730 produce a warning if there is extrapolation beyond the range of the observed y values.

(8) If the output has to be read from a screen, see note (1) to Program 8.1. Recommended places for lines to halt the display are 195–196, 295–296, 505–506, and 585–586.

10.2 Straight line relationships between two random variables

We now look at an entirely different situation, in which two *random variables* are associated with each specimen. A good example, already discussed earlier, is the experiment in which a male student is selected at random and his height in centimetres and weight in kilograms are both recorded. If we took a large enough number of observations we should obtain good estimates of the probability distributions of both heights and weights; this contrasts sharply with the experiment in Section 10.1 where the values of the controlled variable were chosen beforehand.

The statistics for x and for y can nevertheless be calculated in the same way as for the previous experiment, and the formula for finding the best straight line for predicting y from x is unchanged. There is however an equally relevant line, the best straight line for predicting x from y. The latter will be a steeper line, unless all the observed data pairs lie exactly on a straight line. In the extreme case of no correlation between the x and y values, the line of regression of y on x will be the horizontal line $y = \bar{y}$ while the line of regression of x on y will be the vertical line $x = \bar{x}$.

Putting b_Y as the coefficient of x in the former regression line, and b_X as the coefficient of y in the latter regression line, the geometric mean of these two coefficients (with a minus sign if both are negative) is termed the *correlation coefficient* and takes the value 1 or -1 if the observed data pairs lie exactly on a straight line and a value close to zero if there is no significant straight line relationship between the two variables. It is impossible for b_Y and b_X to differ in sign, since they have the same numerator and each has a positive denominator. This numerator may be written as

$$\sum_{i=1}^{n} x_i y_i - \bar{x} \sum_{i=1}^{n} y_i$$

or as

$$\sum_{i=1}^{n} (x_i - \bar{x})(y_i - \bar{y})$$

and when divided by $(n-1)$ it produces the *covariance,* which is clearly analogous to the variances s_X^2 and s_Y^2.

All these results are adequately discussed in most statistics textbooks. However, some textbooks suggest that the regression lines represent the functional relationship between y and x, which is not the case. The regression line of y on x aims to minimize the squared residuals in the 'y direction' so as to optimize the prediction of y from x, and conversely the regression line of x on y minimizes the squared residuals in the 'x direction' and is optimal for predicting x from y.

Still applying the principle of least squares, the 'straight line of best fit' to represent the functional relationship is the line which minimizes the sum of the squares of the distances from the points to the line, i.e. the 'residuals' *perpendicular to the line.* This line is, however, not uniquely defined until we stipulate that the variables must be scaled so that s_Y is allotted the same distance in the y

direction as s_X is allotted in the x direction. It can then be shown that the line is:

$$\frac{y-\bar{y}}{s_Y} = \frac{x-\bar{x}}{s_X} \quad \text{or} \quad \frac{y-\bar{y}}{s_Y} = -\frac{x-\bar{x}}{s_X}$$

depending on whether the correlation coefficient is positive or negative. Unless the two regression lines coincide, the straight line of best fit has an angle of slope intermediate between the slopes of the two regression lines; all three lines pass through the point $(\bar{x},\bar{y})$. If the coefficient of determination is 1 then all three lines coincide; if it is zero or close to zero then the straight line of best fit cannot be determined.

Care should be taken not to muddle the two quite different models discussed in this section and the previous section. It should also be noted that data pairs in which one of the variables is calendar time cannot be regarded as examples of *either* model.

Program 10.2

```
100 REM STRAIGHT LINE RELATIONSHIPS BETWEEN TWO RANDOM
    VARIABLES.
110 EP = 1E-15
120 PRINT: PRINT: PRINT "  CORRELATION AND REGRESSION"
130 PRINT: PRINT "Input number of observations ";
    : READ N: PRINT ;N
140 IF N<2 GOTO 750
150 IF N=2 GOSUB 800
160 IF N>2 AND N<6 GOSUB 900
170 PRINT: PRINT "For each observation, input the values"
180 PRINT "of x and y.": PRINT
200 TX = 0: XX  = 0: TY = 0: YY = 0: XY = 0
210 XL = 1E+11: YL = XL
220 FOR I = 1 TO N
230 PRINT "Input observation No. ";I;": ";
240 READ X: PRINT ;X;" ";: IF X<XL THEN XL = X
250 READ Y: PRINT ;Y: IF Y<YL THEN YL = Y
260 TX = TX+X: XX = XX+X*X: TY = TY+Y: YY = YY+Y*Y
    : XY = XY+X*Y
270 NEXT I
300 MX = TX/N: PRINT: PRINT "Mean of x = ";MX
310 VX = (XX-MX*TX)/(N-1): PRINT "Variance of x = ";VX
320 SX = SQR(VX): PRINT "Standard Deviation of x = ";SX
330 PRINT "Coefficient of Variation of x ";
    : IF XL>0 THEN PRINT "= ";SX/MX
340 IF XL<=0 THEN PRINT "does not exist."
350 MY = TY/N: PRINT: PRINT "Mean of y = ";MY
360 VY = (YY-MY*TY)/(N-1): PRINT "Variance of y = ";VY
370 SY = SQR(VY): PRINT "Standard Deviation of y = ";SY
380 PRINT "Coefficient of Variation of y ";
    : IF YL>0 THEN PRINT "= ";SY/MY
390 IF YL<=0 THEN PRINT "does not exist."
400 CV = (XY-MX*TY)/(N-1): PRINT: PRINT "Covariance = ";CV
410 IF VX<EP OR VY<EP GOTO 850
420 R = CV/SX/SY: PRINT "Correlation Coefficient = ";R
430 PRINT "Coefficient of Determination = ";R*R
440 BY = CV/VX: AY = MY-BY*MX: PRINT
450 PRINT "Regression of Y on X is estimated as:"
460 PRINT "Y = ";AY;" ";: IF BY>=0 THEN PRINT "+ ";
470 PRINT BY;" X"
```

```
480 BX = CV/VY: AX = MX-BX*MY: PRINT
490 PRINT "Regression of X on Y is estimated as:"
500 PRINT "X = ";AX;" ";: IF BX>=0 THEN PRINT "+ ";
510 PRINT BX;" Y": PRINT
520 R = ABS(R): IF R<EP GOTO 120
530 PRINT "Straight line of best fit is:"
540 B = BY/R: A = MY-B*MX
550 PRINT "y = ";A;" ";: IF B>=0 THEN PRINT "+ ";
560 PRINT B;" x": IF ABS(B)<EP GOTO 120
570 PRINT: PRINT "which can be re-arranged as:"
580 PRINT "x = ";MX-MY/B;" ";: IF B>=0 THEN PRINT "+ ";
590 PRINT 1/B;" y": GOTO 120
600 GOTO 120
750 PRINT: PRINT "USELESS!  More observations needed."
760 GOTO 130
800 PRINT: PRINT "With only two observations the line"
810 PRINT "will fit the data exactly.  It will be"
820 PRINT "unreliable for estimation purposes.": RETURN
850  PRINT: PRINT "Correlation Coefficient cannot be
     found.": GOTO 120
900 PRINT: PRINT "With so few observations, the estimated"
910 PRINT "straight lines will be unreliable.": RETURN
950 DATA 6,172,71.2,180,75.3,175,74.9,190,85.1,174,67.4,
    183,73.9
999 END
```

Program notes

(1) Lines 100–390 are similar to the corresponding lines of Program 10.1. The arrays X(255) and Y(255) and the variable YH are omitted since the options to print out the residuals and to predict new *y* values have been omitted from the last part of the program, but they could be added (along with an option to predict new *x* values from given *y* values) if desired. See the notes for Program 10.1.

(2) Lines 400–600 go on to calculate the various results as discussed in the text. Note that additional checks are needed at lines 410 and 560 to ensure that the denominators are not too small for reliable computation.

(3) The variable BY at line 440 is the same as variable B at line 410 of Program 10.1. By writing out on paper the formulae by which these results are arrived at, and also the different ways of obtaining the Coefficient of Determination in the two programs, the reader may improve his understanding of the algebraic relationships as well as of the BASIC.

(4) If the output has to be read from a screen, see note (1) to Program 8.1. Recommended places for lines to halt the display are 195–196, 295–296, 435–436, and 595–596, deleting GOTO 120 from line 590.

10.3 Multilinear regression

Let us suppose that we had carried out the experiment of selecting a male student at random and had recorded his age in years, height in centimetres and weight in kilograms. We could have examined the straight line relationships between each pair of variables using Program 10.2, provided we had performed a sufficient number of repetitions of the experiment. An example of such data (rather sparse) would be:

Age in years	18	19	20	20	18	18
Height (cm)	172	180	175	190	174	183
Weight (kg)	71.2	75.3	74.9	85.1	67.4	73.9

For instance, we find that the straight line regression of weight on height is estimated as: $y' = -60.408 + 0.7544x$, and the correlation coefficient of 0.862 suggests a strong relationship.

However, if our main interest was in trying to estimate weight, we might seek to make use of the information on both age and height for this purpose. Denoting age by x_1 and height by x_2 we would use the principle of least squares to estimate the coefficients in the relationship:

$$\mu_Y(x) = \propto + \beta_1 x_1 + \beta_2 x_2$$

This relationship is termed 'multilinear regression'. Sound BASIC programs for multilinear regression are outside the scope of this text, but we must warn the reader than many unsound BASIC programs are to be found in books and magazines. A common sign of confusion is the use of the term 'independent' to describe the explanatory variables even in cases, such as in the foregoing example, where they are quite obviously not independent. Unsound programs may also produce large errors due to computational problems; good programs will both seek to minimize the errors and print out warnings when ill-conditioning (which we discussed in Section 3.5) is detected.

10.4 Polynomial regression

It was noted in Section 10.1 that a regression relationship may not be a straight line, and that long runs in the signs of the residuals when a straight line is fitted are evidence of a curved relationship. Occasionally, scientific theory may suggest a quadratic curve. In other cases we deliberately fit a quadratic or cubic curve because we know that the y values will have a peak and then fall away, as in the agricultural example we considered.

There are complicated programs for fitting regression polynomials of any degree. As in the case of multilinear regression there are many pitfalls in estimating the curve. Many programs are unsound, and even with a sound program 'a little learning is a dangerous thing'. Polynomials are sometimes appropriate in cases where there are two or more explanatory variables. There are also cases where the appropriate curves are of exponential or trigonometric rather than polynomial form.

A final comment should be made on the use of the term 'linear'. Statisticians use the term 'general linear model' to cover all models in which the estimates of the regression coefficients are obtained by solving sets of simultaneous linear equations. These include polynomial regression curves. We accordingly prefer the term 'straight line regression' rather than 'linear regression' for the model discussed in Section 10.1.

PROBLEMS

(10.1) Examine all the examples and problems in regression and correlation in any textbook in statistics. Do they always make it clear whether the variables are both random variables or whether one is a controlled variable? After deciding which (if any) is appropriate, run each set of data through Program 10.1 or Program 10.2.

(10.2) Find the straight line regression of Y on the controlled variable x for each of the following sets of data:

(a)

x:	1	2	3	4	5	6	7	8	9	10
y:	1	2	4	7	11	15	19	25	31	39

Predict y for $x = 1, 5, 9, 13$.

(b)

x:	0	1	2	3	4	5	6
y:	7	5	5	4	4	3	2

Predict y for $x = 0, 3, 6$.

(c)

x:	2	3	3	5	8
y:	0.5	2.9	2.0	6.1	8.7

Predict y for $x = 0, 10, 100$.

In each of the above examples, examine the residuals and decide whether a regression curve rather than a straight line seems more appropriate. Also decide how much faith ought to be placed in each of the predictions.

(10.3) With experience, one may sometimes apply the methods of Section 10.1 and Program 10.1 to cases where the x variable is not truly a controlled variable even though it is certainly not a random variable. Find the straight line regression of total production cost per week on quantity produced per week in the following example:

Quantity produced in the week (in units of 10 tons)	21	23	26	28	24	25	27	26
Total production cost for the week (in units of £100)	128	133	150	155	137	143	153	145

Predict the total production cost for 20, 30, 40 units per week. Would you find the last of these predictions worth quoting?

(10.4) Use the data in the first paragraph of Section 10.3 to estimate straight line relationships between each of the three pairs of random variables. Note that in practice we are unlikely to be interested in the regression of age on height or the regression of age on weight.

Chapter 11

Other statistical models and techniques

11.1 Goodness-of-fit tests

In a 'goodness-of-fit test', the null hypothesis is that the true probability distribution describing the inherent variability in an observational situation belongs to a particular family of probability distributions. The alternative hypothesis is that it does not belong to that family. This is in contrast to the usual type of test where the form of the distribution is known and the hypotheses are concerned with the value of an unknown parameter.

The commonest form of test is a chi-square test, in which the data are in the form of 'observed frequencies' for a number of classes defined in such a way that the corresponding 'expected frequencies' under the null hypothesis can be computed.

For instance, we may have observed the frequencies of the six possible outcomes in 300 throws of a die. These may be tested for goodness-of-fit to a uniform discrete distribution.

When the observed frequencies and expected frequencies are available, the chi-square test proceeds by treating all the classes as if they were *qualitatively* different; the quantitative ideas which produced the expected frequencies have ceased to be relevant. This principle may be emphasized by making our die a crown-and-anchor die:

Class	*Observed frequency* (F_i)	*Expected frequency* (E_i)
Clubs	43	50
Diamonds	52	50
Hearts	45	50
Spades	56	50
Crown	49	50
Anchor	55	50

The fact that the sum of the observed frequencies is used in calculating the expected frequencies means that there are only five 'degrees of freedom' in the test statistic. It can be shown that, with certain qualifications, the statistic

$$\sum_{i=1}^{NC} \frac{(F_i - E_i)^2}{E_i}$$

has an approximate chi-square distribution with $(NC-1)$ degrees of freedom, where NC is the number of classes, provided that the null hypothesis is true. Departure from the null hypothesis tends to increase the absolute values of $(F_i - E_i)$ and hence the chi-square statistic, and so the test is always an upper-tail test.

For the foregoing example the chi-square statistic works out as 2.8 exactly, with $\nu = 5$, where ν is the number of degrees of freedom. Since the mean of a chi-square distribution is ν, we can accept the null hypothesis without any need to consult the table of the chi-square distribution.

Before considering other examples of goodness-of-fit tests, we shall introduce the modest Program 11.1. This requires the number of classes, observed frequencies, and theoretical probabilities as input. It prints out a warning if the sum of the theoretical probabilities is not close to 1. It sums the observed frequencies and uses the total to calculate the expected frequencies and hence the chi-square statistic. The sum of the frequencies is printed out as a means of checking the input.

Program 11.1

```
100 DIM F(100),P(100)
110 PRINT: PRINT: PRINT "  CHI-SQUARE GOODNESS-OF-FIT TEST"
120 PRINT "(For a fully-specified distribution)"
130 PRINT: PRINT "Input number of classes ";: READ NC
     : PRINT ;NC
140 PRINT: PRINT "Input observed frequency, theoretical
     probability."
150 PRINT: N = 0: T = 0: X = 0
160 FOR I = 1 TO NC
170 PRINT "Class No. ";I;": ";: READ F,P: PRINT ;F,P
180 N = N+F: T = T+P: F(I) = F: P(I) = P
190 NEXT I
200 IF ABS(1-T)>0.01 THEN PRINT: PRINT "WARNING: Total
    probability = ;T
210 FOR I = 1 TO NC
220 E = N*P(I): X = X+(F(I)-E)^2/E
230 NEXT I
240 PRINT: PRINT "Number of classes = ";NC
250 PRINT "Sum of frequencies = ";N
260 PRINT: PRINT "Chi-square = ";X
270 PRINT: PRINT "Degrees of freedom = ";NC-1
280 GOTO 110
300 DATA 6,43,1/6,52,1/6,45,1/6,56,1/6,49,1/6,55,1/6
310 DATA 6,30,1/32,60,5/32,120,5/16,80,5/16,20,5/32,
    10,1/32
320 DATA 11,488,1/36,1021,2/36,1552,3/36,2091,4/36
321 DATA 2629,5/36,2346,6/36,2617,5/36,2095,4/36
322 DATA 1603,3/36,1023,2/36,535,1/36
999 END
```

Program notes

The program should present no difficulties of understanding in the light of the explanations given in the text. If the output has to be read from a screen, lines to halt the display are required as lines 275–276.

For a goodness-of-fit test which might involve a binomial distribution we quote a widely-used textbook. 'Five identical coins were tossed 320 times and the observed frequencies of the number of heads per toss were as given in Table 11.1. Test if the coins are biased at the 5 per cent level.' (Mulholland and Jones, 1968)

Table 11.1

Number of heads	*Frequency*
0	30
1	60
2	120
3	80
4	20
5	10

The authors find that the expected frequencies under the hypothesis of a binomial distribution are respectively 10, 50, 100, 100, 50, 10 when $p = \frac{1}{2}$, compute the chi-square statistic as 68, find that the 5% value for the chi-square distribution with $\nu = 5$ is 11.07, and say 'Therefore, we must reject the null hypothesis that the coins are unbiased'.

The example is completely unsound. This will be obvious if we consider what would have happened if the observed frequencies had been respectively 30, 60, 70, 70, 60, 30. The chi-square statistic would be 102. Does this mean that 'we must reject the null hypothesis that the coins are unbiased' — when the proportion of heads is exactly $\frac{1}{2}$ and so fits the null hypothesis exactly?

The correct test for the null hypothesis stated would be a test of the observed proportion of heads (670 heads in 1600 tosses) against the theoretical proportion of $\frac{1}{2}$ in 1600 observations on a Bernoulli distribution. The chi-square test actually described would be appropriate if the example had read 'Test the hypothesis that the coins are unbiased and the tosses are independent' or 'Test whether the results fit a $b(5, \frac{1}{2})$ distribution'. This again illustrates the conceptual subtleties in statistical inference. It is a minefield of menace for those who do not carefully learn and critically apply the principles; even then, we are all prone to err.

For the simulation experiments in Section 4.5 on the sum of the values thrown on two dice, the expected frequencies (under the null hypothesis that the dice were unbiased and the throws were independent) were respectively 500, 1000, 1500, 2000, 2500, 3000, 2500, 2000, 1500, 1000, 500. The chi-square statistic proves to be 176 for the data obtained using Program 4.2 as modified for the two-dice experiment. Any value over 29.59 when $\nu = 10$ is significant at the 0.1% level, and so the null hypothesis is rejected 'beyond all reasonable doubt'. In actual fact the bad fit arose not because the simulated throws were biased but because successive 'random numbers' were not independent. For the data obtained using Program 4.3 the chi-square statistic proves to be 8.1, and so we accept the null hypothesis.

For our final example we go back to the 'horse-kick' Data 3 in Section 2.5. Statisticians theorized that the data might fit a Poisson distribution. As they were interested only in the type of distribution, and had no theory as to the value of μ, they used the observed mean 0.7 as the estimated mean. The theoretical probabilities for 0, 1, 2, 3, 4, are then 0.4966, 0.3476, 0.1217, 0.0284, 0.0050 respectively, with a total probability 0.0007 for '5 and over'. A requirement for the chi-square approximation is that no class should have an expected frequency of less than 5; this is attained by forming a single class for '3 and over' with theoretical probability 0.0341.

The chi-square statistic, which may be obtained using Program 11.1, works out as 1.96. We did not however have a 'fully-specified distribution'; the fact that μ was estimated from the data means that one degree of freedom is lost.

So for four classes the number of degrees of freedom is 2, and the statistic 1.96 is not significant. We accept the hypothesis that the deaths from horse-kicks are completely random.

If we had wished to test the hypothesis that there is 'accident-proneness', we would have estimated both p and k from the data in fitting a negative binomial distribution as discussed in Section 6.7. Two degrees of freedom would be lost because two parameters were estimated from the data. If the observed variance was smaller than the observed mean we would reject the idea of accident-proneness without attempting to make these estimates.

11.2 Other tests and techniques

In this text we have discussed only a fraction of the recognized statistical tests and techniques. We have concentrated on the basic principles and the BASIC programming, but have also included the simpler and widely-used tests. Brief mention may now be made of the other, more advanced applications of BASIC in statistics.

In 'design models', frequently referred to in books as 'Analysis of Variance', we test the null hypothesis that the means of several populations are equal. This is effectively an extension of the *t*-test as discussed in Section 9.3 where there were only two populations under consideration. There are more complex designs such as two-factor designs, Latin Squares, and many others. There are also special techniques for dealing with the problems of missing values and of outliers which are suspected of being 'rogue readings'.

In goodness-of-fit tests, we could have gone on to consider tests of contingency tables, 'too-good-a-fit tests', and other chi-square tests not based on the 'multinomial' model which underlies Program 11.1. We could also have considered a different type of goodness-of-fit test known as the Kolmogorov–Smirnov test.

There are a number of tests which are variously known as 'nonparametric tests', 'distribution-free tests', or simply 'quick tests'. They include sign tests, rank tests, and run tests. When a computer is available there is no need for 'quick tests' merely to save calculation, but there may be other reasons for wanting to carry out non-parametric tests. In Sections 10.1 and 10.4 we mentioned that runs in the residuals could be an indication that there is a better model than the one being fitted, and it would be possible to add a run test to a polynomial regression program.

Apart from multilinear regression and polynomial regression there are other types of regression relationship, such as trigonometric regression and exponential regression, to be considered. Other statistical techniques which could be applied using BASIC are simple forecasting techniques (often termed 'time series' techniques), the study of 'extreme value' problems, and the methods of directional statistics.

Fields of application sometimes require their own distinctive probabilistic models, such as the hyperexponential and Weibull distributions in reliability theory. There are special techniques for decision-making in the presence of uncertainty; sequential sample schemes in quality control have suggested sequential decision-making procedures in managerial fields such as investment evaluation.

All these tests and techniques can be programmed in BASIC. We have not discussed the more advanced techniques of multivariate analysis, cluster analysis, spectral techniques in time series, etc., since the complex programs necessary to make full use of these powerful statistical procedures are more suitably tackled in FORTRAN, Pascal, or even more specialist mathematical languages. But none of them can be pursued without a proper understanding of basic statistics.

PROBLEMS

(11.1) Test the goodness-of-fit to a uniform discrete distribution of the following results of 720 throws of a die:

Value thrown	*Observed frequency*
1	112
2	104
3	134
4	121
5	125
6	124

(11.2) In Section 4.5 the results of simulation runs each of length 18 000 were given for two methods of generating the sum of the values thrown on two dice, and reference was made in Section 11.1 to goodness-of-fit tests on these two sets of results. Use Program 11.1 to carry out these tests.

Two further simulation methods were also tried. In method 3, the pseudo-random-number generating technique using $M = 8192$, $D = 67\ 101\ 323$, was applied by direct arithmetic, as if the BASIC could perform exact integer arithmetic using these values of M and D (as a FORTRAN or Pascal compiler normally would do). In method 4, the same technique was applied but the arithmetic was performed as in Program 4.4 to ensure that it would be exact. The two sets of results were:

Sum of values	*Method 3 frequencies*	*Method 4 frequencies*
2	547	462
3	1071	1017
4	1625	1545
5	1848	1983
6	2620	2483
7	3039	2962
8	2393	2580
9	1696	2014
10	1779	1522
11	1008	980
12	374	452

Use Program 11.1 to test the goodness-of-fit of these two further methods, the expected frequencies being as specified in Section 11.1.

(11.3) Fit a negative binomial distribution to the following data, estimating both p and k from the data as discussed in Section 6.7.

Number of accidents	*Frequency*
0	25
1	25
2	17
3	12
4	8
5	7
6	4
7	2

Calculate the theoretical probabilities (using Program 6.5 if available) and multiply them by 100 to obtain the expected frequencies. For the chi-square goodness-of-fit test it is necessary to amalgamate '6 and over' into a single class, making 7 classes in all. Program 11.1 may then be used to calculate the chi-square statistic, but as the theoretical distribution was not 'fully-specified' (it had two parameters estimated from the data) the number of degrees of freedom for the 7 classes will be 4 and not 6 as printed out by the program.

(11.8) Try to develop Program 11.1 so that it will accept data such as the 'horse-kick' data to be fitted to a Poisson distribution, use the mean of the data as the estimate of μ in calculating the theoretical probabilities and hence the expected frequencies by the method of Program 6.6, compute the chi-square statistic after amalgamating classes as necessary, and print this out with the correct number of degrees of freedom. Try to develop it further so that it will similarly produce the correct chi-square statistic and correct number of degrees of freedom when fitting a negative binomial distribution.

(11.9) Use Program 11.1, with such developments as you have been able to incorporate, for all examples of chi-square goodness-of-fit tests in your statistics textbook with which it can cope.

Index

References to 'computers', 'programming', 'programs', 'statisticians' and 'statistics' are too numerous to be listed, but there are entries for 'mainframe computers', 'microcomputers', 'programming languages', and 'statistic'. For common BASIC words such as PRINT, only the first six references and any particularly important references are listed; this is also the policy for 'events', 'observations', 'parameters', 'populations', 'probability', 'randomness', 'samples', 'specimens', 'statistic' and 'unbiasedness'.